The Ultimate Security Survey

Second Edition

The Ultimate Security Survey

Second Edition

James L. Schaub, CPP

and

Ken D. Biery, Jr., CPP

Butterworth Heinemann

Boston • Oxford • Johannesburg • Melbourne • New Delhi • Singapore

THE ULTIMATE SECURITY SURVEY, SECOND EDITION

James L. Schaub, CPP, Ken D. Biery, Jr., CPP

A member of the Reed Elsevier group

Recognizing the importance of preserving what has been written, it is the policy of Butterworth–Heinemann to have the books it publishes printed on acid-free paper, and we exert our best efforts to that end.

Butterworth–Heinemann supports the efforts of American Forests and the Global ReLeaf program in its campaign for the betterment of trees, forests, and our environment.

Library of Congress Cataloging-in-Publication Data
Schaub, James L.
The ultimate security survey / James L. Schaub, Ken D. Biery. — 2nd ed.
p. cm.
ISBN 0-7506-7091-6 (alk. paper)
1. Industries—Security measures. 2. Commercial buildings—Security measures. 3. Industrial safety. 4. Retail trade—Security measures. 5. Computer security. I. Biery, Ken D. II. Title.
HD61.5.S33 1998
658.4'73—dc21
98-20602
CIP

British Library Cataloguing-in-Publication Data
A catalogue record for this book is available from the British Library.

The Publisher offers special discounts on bulk orders of this book.
For information, please contact:
Manager of Special Sales
Butterworth–Heinemann
225 Wildwood Avenue
Woburn, MA 01801-2041
Tel: 781-904-2500
Fax: 781-904-2620

For information on all Butterworth–Heinemann publications available, contact our World Wide Web home page at: http://www.bh.com

10 9 8 7 6 5 4 3 2 1

Printed in the United States of America

CONTENTS

FOREWORD

INTRODUCTION

Over the years we have conducted hundreds of Security Surveys and Risk Assessments, many of them for a variety of reasons. We were first introduced to the Security Survey at the National Crime Prevention Institute, then at the Boston Chapter of ASIS annual EXPO Security Program. We found the security survey to be a very effective Crime Prevention/Loss Prevention tool.

However, you must understand the psychology of this survey report which you are about to file. You will be making recommendations to improve security, and you will want your recommendations implemented. Therefore, we suggest that you evaluate the results of every survey you conduct, in order to be more effective with your next survey.

DEFINITION

A security survey is "a critical on-site examination and analysis of an industrial plant, business, home, public or private institution, to ascertain the present security status, to identify deficiencies or excesses, to determine the protection needed, and to make recommendations to improve the overall security." *1

It is interesting to note that a definition of crime prevention as outlined by the British Home Crime Prevention Program — "the anticipation, recognition and appraisal of a crime risk and the initiation of action to remove or reduce it" — could, in fact, be an excellent definition of a security survey. The only difference is that a survey generally does not become the "action" as such but rather a basis for recommendations for action.

This definition can be divided into five component parts and analyzed so that its implications can be applied to the development of a working foundation for the security surveyor.

1. The Anticipation. How does the anticipation of a crime risk become important to the security or crime prevention surveyor? Obviously, one of the primary objectives to a survey is the anticipation or prevention aspects of a given situation. Thus, an individual who keeps anticipation in the proper perspective would be maintaining a proper balance in the total spectrum of security surveying. In other words, the anticipatory stage could be considered a prognosis of further action.

2. The Recognition. What means will provide an individual who is conducting a survey of the relationships between anticipation and appraisal? Primarily, the ability to recognize and interpret what seems to be a crime risk becomes one of the important skills a security surveyor acquires and develops.

3. The Appraisal. The responsibility to develop, suggest, and communicate recommendations is certainly a hallmark of any security survey.

4. The Criminal Risk. This, as defined in this text, is the opportunity gained from crime. The total elimination of opportunity is the most difficult, if not the most improbable, task. Thus, the cost of protection is measured in:

   a. Protection in depth.

   b. Delay time.

   Obviously, the implementation of the recommendation should not exceed the total (original/ replacement) cost of the item(s) to be protected. An exception to this rule would be human life.

5. The Initiation of Action to Remove or Reduce the Risk Of Crime. This section indicates the phase of a survey in which the recipient of the recommendations will make a decision to act, based on the suggestions (recommendations) set forth by the surveyor. In some cases, the identification of a security risk is made early in a survey, and it is advisable to act upon the recommendation prior to the completion of the survey.

The responsibility to initiate action based on recommendations is the sole duty of the recipient of the survey. This is to suggest that the individual who receives the final evaluation and survey will be the individual who has commensurate responsibility and the authority to act.

There are basically three types of surveys:

1. Building Inspection is advising a tenant in a large complex of his vulnerabilities as they pertain to the physical characteristics of the dwelling.

2. A Security Survey, on the other hand, would be conducted on the whole complex versus doing only a portion of the site.

3. A Security Analysis is more of an in-depth study including risk management, analyzing risk factors, environmental and physiological security measures, analyzing crime patterns, and fraud and internal theft.

THE BEST TIME TO CONDUCT THE SURVEY

Most crime prevention officers and security directors agree that a survey is most effective:

1. After a crisis within the corporation.

2. After a breaking and entering or major larceny.

3. Upon request.

There are times when a merchant hears he can get something for nothing and thereby calls the crime prevention officer in the town to conduct such a survey, but in reality has no intention of spending a dime for improvement. A close friend of mine conducted a detailed security survey on a factory warehouse and office building. The recipient of the survey followed only one of his recommendations, which was to leave a light on over the safe in the back room of his warehouse. The owner had completely disregarded other recommendations such as hardware improvements on doors, windows, and skylights. Unfortunately, thieves returned and almost put him out of business.

When asked to conduct a survey, the first question you should seek to answer is "Why are we here?"

CLASSIFICATIONS OF SURVEY RECOMMENDATIONS

The various classifications of recommendations can be best explained through an example. The classifications are maximum, medium, and minimum. The example selected is a museum that contains $250 million in various art treasures; the complex has no security.

Maximum Security

Obviously the museum needs an alarm system; therefore, to apply our maximum security classification recommendation, it should read:

Alarm the perimeter (all exterior and interior doors, all windows and skylights). Four panic alarms should be installed at various locations and six paintings which are worth 12 million dollars should be alarmed — each alarm on a separate 24-hour zone.

I specifically did not mention ultra-maximum security because this term applies to an armed camp — machine guns, men in full battle dress (guards) armed with semiautomatic rifles, grenades, flame throwers, mines and locking devices equipped with dynamite which will blow up when the intruder attempts to pick the lock. It is dramatic and it is ultra-maximum. It is not ridiculous for Fort Knox to provide ultra-maximum security to protect its billions in gold bullion.

Medium Security

A medium-security classification recommendation would read:

Alarm all basement windows and all ground floor windows, which are at the rear of the building. Install one panic alarm by the main entrance. Alarm each of the six paintings worth 12 million dollars on a separate 24-hour zone.

Minimum Security

Finally, a minimum-security classification recommendation would read:

From a risk management point of view, alarm the six paintings, which are worth 12 million dollars, each painting to be alarmed on a separate 24-hour zone.

Zero Security

This classification speaks for itself.

FIRST STEP

These three examples clearly show the degree of security one can obtain by trying to plan a security package. I have stated these examples because your first step in conducting a security survey is the interview you have with the individual to whom you turn over your report. It is during this interview that you form an appraisal on the degree of protection, which is required and needed.

There are times when you may have to state all three recommendations in a report. There are also times when you must be conscious of the fact that you may force the receiver of your report to accept less security than you suggested because you did not thoroughly and clearly explain your security points.

After your report has been typed, go over it several times to ensure that it is worded the way you want it presented.

DEVELOPING YOUR SECURITY POINTS

Like most professionals, we need tools to do an effective job. The following are suggested to assist you when conducting your surveys: tape measure, floor plans, magnifying glass, flashlight, camera with flash, small tape recorder, screwdriver, penknife, pencil, and paper.

Your survey is to be conducted systematically so the recipient can follow your recommendations in some kind of order. Start with the perimeter of the building. Once inside the building, start at the basement and work your way to the attic. Do not be afraid to be critical of the area that you are in. This is what the recipient wants.

After you have done several surveys you will develop a style of putting them together and they become easy.

DOS AND DON'TS IN DEVELOPING YOUR REPORT

Dos:

1. Be honest in your recommendations. You are the expert!
2. Call the shots as you see them.
3. Be critical — physically tear the property apart.

Don'ts:

1. Don't over exaggerate your reports. They are too important.
2. Don't inflate the report with maps and floor plans.
3. Don't repeat your statements.

The written report should include the following:

Page One:

Introduction or sample cover letter.

Page Two:

1. Identification of the building or facility.
2. Specific statement of the major problem.
3. Alternative recommendation to the problems.
4. List of your further recommendations.

General statements such as the following can be included in the report:

> Complete a physical inventory of all property at least annually. Your inventory should have listed the name of the item, the manufacturer, model, serial number, value, description, and date of acquisition.

OTHER KEYS TO BEING AN EFFECTIVE SURVEYOR

It is only when you have developed the ability to visualize the potential for criminal activity that you will become an effective crime scene surveyor. This ability is the part of the process that is referred to as an art. Nonetheless, it is important that when you arrive on a survey site you are prepared to give a property owner sound advice on the type of security precautions to consider.

In summary, to be a good crime prevention surveyor, you will have to be a good investigator. You must understand criminal methods of operation and the limitations of standard security devices. In addition, you must be knowledgeable about the type of security hardware necessary to provide various degrees of protection.[2]

Do not underestimate the above statement! Physical security design and application is a complicated science with the rapid changes in technology. For example, you have recommended that closed circuit television be installed. But you did not specify the type of chip cameras, method of transmission, fiber optics, microwave, laser, digitized, or time-lapse recorder.

To be effective you must also consider the cost factor of your recommendations.

TEN POINTS OF SECURITY CONCERN

1. General purpose of the building (i.e., residence, classroom, office, etc). Consider the hours of use, people who use the building, general hours of use, people who have access, key control, maintenance schedule. Who is responsible for the maintenance schedule? Is the building used for public events? If so, what type and how often? Is the building normally opened to the public? Identify the significant factors and make recommendations.

2. Hazards involving the building or its occupants. List and prioritize, i.e., the theft of office equipment, a wallet, or items from stockrooms. Also, identify potential hazards which might pose problems in the future.

3. Police or security guard applications. What can they do to improve the response to the building and occupants from a patrol, investigation, or crime prevention standpoint? Would the applicant of guards be operationally effective and/or cost effective?

4. Physical recommendations. Inspect doors, windows, lighting, and access points. Recommend physical changes which would make the building more secure, such as pinning hinges on doors and fences.

5. Locking equipment. Make specific recommendations. Locks, equipment to be bolted down, potential application of card control and key control devices.

6. Alarms. Would an alarm system be cost effective? Would the use of the building preclude the use of an alarm? Are the potential benefits of an alarm such that the building use should be changed to facilitate the use of an alarm? Consider all types of alarms, building-wide or specific offices. Consider closed circuit television and applications for portable or temporary alarm devices.

7. Storage. Are there specific storage problems in the building, i.e., expensive items which should be given special attention, petty cash, stamps, calculators, microscopes? Make specific recommendations.

8. Signage. Are there adequate "No Trespassing" signs posted? Are other signs needed?

9. Custodians. Can they be used in a manner which would be better from a security standpoint?

10. Standards. Does the site comply with established security standards, community standards, and state or county requirements?

PERSONALITY OF THE COMPLEX YOU ARE SURVEYING

Each complex that you survey will have a distinctive personality. Let us take an average building, which is open from 9 A.M. to 5 P.M. The traffic flow is heaviest during this period. From 5 P.M. to 1 A.M., the building is closed to the public. Some staff members may work late. Who secures the building? At 1 A.M., the cleaning crew arrives and prepares the building for another day. The whole personality of the complex must be taken into consideration before your report is completed.

Let us take a further example of building personality. The complex is 100 feet by 100 feet, and it has two solid core doors, one large window at the front of the building, and is air conditioned.

Case 1: The complex is a credit union on the main street directly next door to the local police department versus the same credit union on the edge of town.

Case 2: This is a large doctor's office. The doctor is an art buff and has half a million dollars in art in the office versus a doctor who has no art but does have a small safe that contains about $200 worth of Class A narcotics inside.

Case 3: This building houses a variety store which closes at 6 P.M. versus a liquor store which is open until 2 A.M. on both Saturday and Sunday.

In the above three cases, I have given six examples of the personality of a complex. As I have stated, your recommendations must be tailored to fit the lifestyle and vulnerabilities of these buildings.

POSITIVE AND NEGATIVE ASPECTS OF MAKING RECOMMENDATIONS

In making your recommendations for security improvements, you must consider the consequences of your suggestions in the event that the property owner implements it. There are negative as well as positive aspects involved.

Take, for example, a housing complex that has a high crime rate from outsiders and within. Your recommendation is; "Build a 10-foot high fence around the complex."

POSITIVE ASPECTS

The reduction of crime — the environment can be designed so that the individual considering the criminal act feels that there is a good chance for him to be seen by someone who will take action on his own to call the police.

Vandalism will be less — the target of attack can be made to appear so formidable that the person does not feel able to reach the target. It will add to the physical aesthetics of the area through environmental design.

Visual negative impact — this insures the property of the residence adding to their secure environment. Limiting the number of points of entry and establishing access control will primarily decrease crime opportunity and operate to keep unauthorized persons out.

NEGATIVE ASPECTS

A fortress environment may create more of a psychological barrier than a physical one, and it is socially undesirable.

CRIME ANALYSIS

Although it is not necessary for you to be a statistician, the more you do know about and understand the local crime problems, the better equipped you will be to analyze the potential crime risk loss when surveying a business or a home.

Crime analysis collection is the gathering of raw data concerning reported crimes and known offenders. Generally, such information comes from crime reports, arrest reports, and police contact cards. This is not to say that these are the only sources available to collect crime data.

The analysis process applied to criminal activity is a specific step-by-step sequence of five interconnected functions:

1. Crime data collection
2. Crime data collation
3. Data analysis
4. Dissemination of analysis reports
5. Feedback and evaluation of crime data

Crime analysis of the site that you are surveying will supply you with specific information to help you further harden the target in specific areas where losses have occurred. It is a way to respond "after the fact" when a crime was committed.

KEY CONTROL

Key control is a very important factor in conducting a survey. Check to find out if the owner or manager picks up keys from employees at their termination. Do they have an accurate record of who has which keys? Within a few short minutes, you should realize whether or not the recipient of your survey has a problem.

Almost every company has some sort of master key system. The reason is that many people must have access to the building without the inconvenience of carrying two dozen keys around every day. Master keys are required for company executives, middle managers, security department, as well as the maintenance department.

CLOSED CIRCUIT TELEVISION

Closed circuit television (CCTV) is a valuable asset to any security package and an even more valuable tool if hooked up to a recorder. CCTV is a surveillance tool which provides an added set of eyes. If this equipment is on the site you are surveying, it is your responsibility to evaluate is operation and effectiveness.

1. Is it working properly?

2. How is it being monitored?[3] Is it statistically placed where it will be most beneficial?[4] What is the type and quality of the lens and components?

INTRUSION ALARMS

If the site you are surveying already has an alarm system survey it completely. Physically walk through every motion detector unit. Evaluate the quality of the existing alarm products versus what is available to meet the needs of the client.

We surveyed a warehouse recently which was only five years old. It was interesting to note that the warehouse had a two-zone alarm system. The control panel was to the right of the front door, which was about 15 feet from the receptionist. Both alarm keys were in the key cylinders and according to the president of the company, "The keys have been there since the system was installed." My point is, for a dollar, another key could be duplicated and now the area is vulnerable to attack.

While doing a survey of an art gallery in New York, the security director stated that he had not had a service call on his alarm system in two years. We then proceeded to physically check every motion detection unit and magnetic contact. You can imagine his reaction when he found out that 12 of the 18 motion detection units were not working.

In conclusion, intrusion alarms come in all shapes and sizes using a variety of electronic equipment. It is advisable to be familiar with the state of art of electronics so that you can produce an effective report.

LIGHTING AND SECURITY

What would happen if we shut off all the lights at night? Stop and think about it!

The results of such a foolish act would create an unsafe environment. Senior citizens would never go out, and communities would have an immediate outbreak of thefts and vandalism. Commercial areas would be burglarized at an uncontrollable rate. Therefore, lighting and security go hand in hand.

The above example may seem farfetched, but in fact installation of improved lighting in a number of cities has resulted in the following:

1. Decrease in vandalism.
2. Decrease in street crimes.
3. Decrease in suspicious persons.
4. Decrease in commercial burglaries.
5. In general, a reduction in crime.

When discussing lighting in your report, consider established standards which will make your report stronger.

STREETLIGHTS

Streetlights have received the most widespread notoriety for their value in reducing crime. Generally, streetlights are rated by the size of the lamp and the characteristics of the light dispersed. Four types of lighting units are utilized in street lighting. The most common, and oldest, is the incandescent lamp. It is the least expensive in terms of energy consumed and the number needed. As a result, incandescent lighting is generally recognized as the least efficient and economical type of street lighting for use today.

The second type of lighting unit that, as a recently developed system, has been acclaimed by some police officials as "the best source available," is the high-intensity sodium vapor lamp. This lamp produces more lumens per watt than most other types. It is brighter, cheaper to maintain, and the color rendition is close to that of natural daylight.

The third and fourth types of devices commonly used for street lighting are the mercury vapor and metal halide lamps. Both are bright and emit a good color rendition. However, the current trend is to use high-intensity sodium vapor lamps because they are more efficient than mercury vapor lamps.

OTHER SECURITY ASPECTS

Depending on the type of facility you are surveying, the following should be reviewed:

1. Communications network, walkie-talkies, and locations of interior and exterior phones.

2. Guard force and security personnel, their training, police powers, uniforms, use of badges, and method of operation.

Your objectives are to identify vulnerabilities, evaluate the site, and provide critical assessment. Methodology and style is purely that of the surveyor, but do not forget it also represents a document from you and your department.

SECURITY SURVEY FOLLOW-UP

The follow-up to your security survey takes many forms, from actually sitting down with the recipient to going by the site and seeing if any changes have actually occurred. Some police departments conduct five to seven surveys per day. They do not evaluate their performance because of the time and manpower involved. In this way, they are failing to examine their own effectiveness. The reason for the follow-up is to encourage increased compliance and to ensure that recommendations are understood. Without this step, you will not know if the recipient has taken any action.

The basic security survey framework consists of five steps:

1. Generating the survey request
2. Conducting the physical inspection
3. Delivering survey findings and recommendations
4. Following-up after the report is completed
5. Evaluating the program

For every crime, there is a crime prevention or loss reduction defense or procedure which, if followed, could delay or deny a criminal from committing that act.

Physical security is implementing those measures to delay or deny unauthorized entry, larceny, sabotage, fire, and vandalism. This chapter of security surveys is designed to help both private security and public law enforcement harden a target and to provide assistance to the community to further reduce losses.

CONCLUSION

Remember that your survey has several objectives — one of which is to reduce risk and reduce losses. If a risk cannot be avoided without creating several operational difficulties, the next step is to reduce it to the lowest level which is compatible with the client's operation.

All of this is accomplished by reviewing documents, inspecting the site, interviewing key personnel, and formalizing your report.

The Security Survey or Risk Assessment is a management tool designed to help a complex meet these objectives:

1. The identification of assets.
2. The protection needed.
3. The probability of risk.
4. The past and future losses of the complex.
5. The potential impact of losses and risk.
6. The cost of recommendations.
7. The overall geographical location.
8. The structure of the complex.
9. The operational aspect of the complex.
10. The recommendations.

"Risk analysis/assessment is the process used to establish priorities to protect corporate assets."*[6]

**150 Things You Should Know About Security.*

Louis A. Tyska, CPP
Lawrence J. Fennelly, CPO

REFERENCES

1. Momboisse, Raymond M. *Industrial Security for Strikes, Riots and Disasters.* Springfield, IL: Charles C Thomas, 1968, p. 13.
2. Kingsbury, Arthur A. *Introduction to Security and Crime Prevention Surveys.* Springfield, IL: Charles C Thomas, pp. 6, 7.
3. Washington Crime Watch. *Crime Prevention Training Manual.* Security Survey Section, p. 8.
4. National Crime Prevention Institute. *Understanding Crime Prevention*, Boston: Butterworth-Heinemann, p. 49.
5. Fennelly, Lawrence J. CPO. *Handbook of Loss Prevention and Crime Prevention*, 3rd ed. Boston: Butterworth-Heinemann, 1996, p. 33.
6. Tyska, Louis A. CPP, and Fennelly, Lawrence J. CPO, *150 Things You Should Know About Security*, Boston: Butterworth-Heinemann, 1998.

ACKNOWLEDGMENTS

The authors would like to thank the following people.

Larry Fennelly and Lou Tyska for their continued assistance in providing guidance and material for our projects.

Fred Zagurski, of Zagurski and Associates, for supplying much of the information in the appendices. Mr. Zagurski has been a long-time information provider for our projects, as well as a good friend.

Byron B. Thatcher, Major, USAF, AFIWC/EACA for supplying the information used in Appendix A — Glossary of Terms. We have attempted to reach Mr.Thatcher. Unfortunately, he moved on from his position after posting his version of the Glossary of Terms on the World Wide Web.

And last, but certainly not least, our wives — Leslie Schaub and Kymberly Biery. Without their ongoing support and the fact that they look the other way when it comes to computer widowhood, we would never have completed our books.

FACILITY INFORMATION

Survey-control number: ______________________ Date: ______________________

Name of facility/site: ______________________

Company Name: ______________________

Address: ______________________

Surveyor/Consultant: ______________________

Time spent on project: ______________________

Project Notes:

____

____

____

____

____

____

Directors:

______________________ ______________________

______________________ ______________________

______________________ ______________________

______________________ ______________________

______________________ ______________________

Managers:

______________________ ______________________

______________________ ______________________

______________________ ______________________

______________________ ______________________

______________________ ______________________

______________________ ______________________

Officers:

Facility contacts and telephone numbers:

Main facility telephone numbers:

Computer system operator's telephone numbers:

Emergency telephone numbers:

Name	Position	Work	Home

General purpose of site:

Facility's range of hours in use:

Area	From	To

High activity use (hours/days):

Facility ownership:

Building:	Owned	Leased	Rented
Property:	Owned	Leased	Rented

List other tenants in building(s):

____

____

____

____

____

____

Number of people who have access to the facility:

____

____

____

____

____

____

Number of individuals who have access to critical areas, description of duties, names, (computer areas, telecommunication areas, executive offices, critical maintenance areas, etc.):

____

____

____

____

____

____

Is the site normally open to the public? Y/N

Number of buildings, rooms, or offices occupied by the facility:

Who performs facility maintenance?

Maintenance schedule:

Estimated dollar value of equipment and property in critical facility areas:

Location of areas with highest dollar value:

Location of areas containing sensitive or competitive information:

Site description:

Obvious security problems or vulnerabilities (If applicable):

____

____

____

____

____

Attach the following drawings, sketches, plans, or schematics (once attached this document should become company proprietary information and should be secured accordingly):

- Facility building and computer area locations
- Facility and/or building area ingress/egress points
- Facility perimeter drawings
- Locations of doors, windows, and similar openings
- Critical or expensive asset locations
- Vulnerable assets or merchandise locations
- Local Area Network wiring diagrams
- Computer area ingress/egress points
- Critical computer systems locations
- Media storage locations
- Alarm placement and diagrams (schematics)
- Evacuation plan drawings
- The last security site survey conducted
- Disaster recovery plan for all computing and telecommunication assets
- Cabling records
- Terminal connections records
- Color coding records
- Conduit routing diagrams

PROJECT MANAGEMENT

MEASURING OVERALL PROJECT EXPECTATIONS

1. What are the project expectations?

   A. Are scope and objectives well defined?

   B. Are roles and responsibilities understood?

   C. Have deliverables, and results to be produced, with acceptance criteria been established?

   D. Are resources and information needed to ensure successful project completion available?

   E. Have the responsibilities, support, and involvement required of the customer been clearly stated?

2. Who sets the project expectations?

3. What are the project requirements as per the contract?

4. List any services provided prior to contract signing.

5. List any dates for project milestones and their corresponding outputs.

6. What are the priorities for

   A. Technical performance?

B. Schedule?

C. Cost?

7. What final action constitutes contract completion and successful delivery?

PROJECT START-UP CHECKLIST

8. Is project mission "need" or "opportunity" defined?

9. Does the client have enough budget available to cover the cost of the contract?

10. List the negotiated budget amount and codicils (i.e., travel, equipment, etc.).

11. Are travel expenses going to accrue in this project?

12. Are any contractual agreements needed for this project?

13. Is the project mission related to the strategic plan or a program mission?

14. Are the preliminary functions and requirements properly identified?

15. Is a preliminary schedule, including an anticipated start date, identified?

16. Is there a rough order-of-magnitude estimate or exact final cost required for this?

17. Is there an estimated cost outlined for each contract phase?

18. Have you outlined anticipated real costs to you for each contractual phase?

19. Is there any physical risk to you in performing this contract (i.e., the need to access high or hazardous areas)?

20. Have control management and records management been addressed for this project?

21. Has a comprehensive project description been developed?

22. Has a Project Support Statement of Work and rules of engagement been defined?

23. Have specific products and services which support your involvement in the project been defined? (including delivery dates)?

24. Have any special requirements been established?

25. Have all necessary approvals been rendered?

PROJECT RECORDS RETENTION

26. Are concurrent projects recorded separately?

27. Project name

28. Do these project records include the following information for each client:

    A. Client

    B. Contact

    C. Phone

    D. Address

    E. City / State / Zip

29. Facility location (address, directions, etc.)

30. Description

31. Start date

32. Milestone dates

33. Anticipated project completion date

34. Actual project completion date

35. Project length (i.e., long term or short term)

36. Status

37. Facility site

38. Building

39. Room

40. Organization

41. Mail address

42. Mail stop

43. Project plans

44. Job or project number

45. Project phases

46. Project phase costs

47. Follow-ups/action items required

48. Background notes

49. Purposes/objectives of the projects

50. Benefits (both real and perceived) to the customer

51. People involved in this project:

    A. Name

    B. Phone

    C. Fax

    D. Address

    E. E-mail

    F. Pager number

    G. Comments

    H. Function to the project (customer employee, vendor, etc.)

PROJECT STRATEGY DEVELOPMENT

52. What work activities do you need to accomplish the project? (List activities.)

53. In what order do you need to accomplish them? (Create logical list of steps.)

54. How long will each activity take? (List duration.)

55. What do you need from others to complete each task?

56. What are your deliverables for each step or phase and to whom are they delivered to?

57. Does the Work Breakdown Structure accurately reflect the work being done?

58. Are all major milestones being accomplished on schedule?

59. Are cost projections consistent with the monthly financial projections?

TECHNICAL REVIEWS

60. Are technical reviews required to properly perform the project?

61. Who should be involved in the technical reviews?

62. When should these reviews be held?

63. Are technical reviews documented with means for following up assigned action items?

TRAINING

64. Is customer training required to complete the project?

65. What type of training is involved?

66. Does this training require any special equipment?

67. Identify who needs the training and when as a milestone in the project.

68. Identify those individuals who have not discharged their training adequately.

69. Have all customer employees who are required to fulfill the project demands been properly briefed?

70. Have all reportable occurrences been properly documented?

TEAM ASSESSMENT

71. Describe each team associated with the project.

72. Who is involved in each team (names, phone numbers, etc.)?

73. What is the scope of each team?

74. What limitations are placed on each team?

75. What decisions have already been made, before the first team meeting which will affect the team's outcome?

CUSTOMER-SIDE DISENGAGEMENT CHECKLIST

76. Are all contracts concluded?

77. Are final cost reports available?

78. Have all customer feedback reports been provided?

79. Have all acceptance criteria been met?

80. Have all approvals been received?

81. Have all obligations been met?

82. Have all issues and action items been resolved?

CONSULTANT-SIDE DISENGAGEMENT CHECKLIST

83. Has a final review been held?

84. Have project results been documented?

85. Have products and services been delivered and accepted?

86. Has responsibility for products and services been transferred?

87. Has customer feedback been obtained?

88. Has post-project customer service been negotiated?

89. Have all contracts been closed out?

90. Has there been a financial close out?

91. Have process improvements been identified and documented?

92. Has a post-project analysis been performed and documented?

93. Have historic records been archived?

94. Have resources been released or redirected?

95. Has the project team been reassigned?

96. Have notes of appreciation been sent to all team members?

PROTECTING INFORMATION AND INFORMATION SYSTEMS

BACKUP AND INFORMATION STORAGE PROTECTION

1. Is information backed-up at least once a week (or more often if stated in the written computer security policy)?

2. Who is responsible for performing the backup of the computer system?

3. Who performs critical information backups?

4. Where is information and commercial software stored when not in use?

5. Are all diskette and tape storage areas free of strong magnetic fields?

6. Are all diskette and tape storage areas free of excessive moisture?

7. Are all diskette and tape storage areas free of excessive heat sources?

8. Are all diskettes and tapes that are removed from the storage area required to be signed out on a tracking log?

9. Does the storage area's tracking log require the diskettes and tapes returned by a certain date?

10. If these materials are overdue, are they immediately located?

11. Are all work diskettes and tapes used for processing during each shift inventoried daily?

12. Are all newly received diskettes and tapes immediately entered into the inventory?

13. Are all work diskettes and tapes properly labeled?

14. Are all work diskettes and tapes allocated from the available inventory?

15. Are there procedures for tracking work diskettes and tapes used for critical processing needs or programs?

16. Do these procedures also mandate the return of all stock which is not used for permanent storage?

17. Are there procedures which would prevent any lost or use of work diskettes and tapes without being detected?

18. Does the facility utilize an Electronic Asset Protection (EAP) System for tracking magnetic storage media?

19. If so, have EAP tags been affixed to all computer diskettes and storage tapes and has the facility considered purchasing diskettes manufactured with these tags already installed?

20. Have EAP sensors been located at each entry and exit point of the backup and information storage areas?

21. Is the EAP system integrated with a closed circuit television (CCTV) system?

22. Is the EAP system integrated with the facility's intrusion alarm system?

23. Is the EAP system integrated with the facility's access control system?

24. Are all of these systems integrated in a configuration to allow them to be monitored simultaneously from a single console?

25. Are there procedures for thoroughly erasing all work diskettes and tapes rather than simply overwriting them with new data?

26. Are there procedures for tracking all work diskettes and tapes used by programmers?

27. Are there procedures for removing all work diskettes and tapes from inventory which are no longer used or useable?

28. Where is information and commercial software stored when not in use?

COMPUTER AREA ACCESS CONTROL

29. Are there periodic computer security reviews to identify threats and vulnerabilities?

30. When are these reviews conducted?

31. Who actually performs the computer security reviews?

32. Do building entry points such as equipment doors, maintenance entrances, and emergency exits, utilize some form of access control?

33. Does the computer area have controls that are separate from the building or computer room access control systems?

34. Is computer area entry controlled at all times? (If not, when is it not controlled?)

35. Is computer area entry controlled when the computer itself is unattended?

36. Is there an automated access control system implemented to limit accessibility to the computer system?

37. Are computer area entry points controlled by access control readers?

38. How many individuals have magnetic cards, badges, or proximity cards permitting entry to the computer area?

39. When a person's computer area access is taken away, are the appropriate updates made to access control systems and authorization lists to prevent entry to this area?

40. When a person's computer area authorization is revoked, are authorization lists revised, locks/combinations changed, and badges, keys, and cards surrendered?

41. Is this denial process efficient enough to prevent unauthorized individuals from damaging computer area resources?

42. What are the procedures permitting computer area access to emergency personnel in case of fire, major power outage, or other emergency or disaster?

43. If an automated system is utilized, does it record employee entry/exit time, identification of entrant, and authorization mechanism (badge, card, etc.)?

44. Does computer area CCTV and alarms transmit to a location where timely responses or other actions may be taken?

45. Where do these monitors and alarms transmit (main guard station off-site, guard station in another building, guard station in the same building, outside vendor monitoring company)?

46. Are there written guidelines for evaluating emergency responses to CCTV systems and/or alarms? (If so, what are they?)

47. How is computer area entry controlled during emergency situations?

48. Have computer area emergency exits been configured so they are only capable of being operated from the inside?

49. Are emergency exits alarmed?

50. Are there any physical access control systems limiting access to the computer areas?

51. Are computer area entry doors controlled by cipher lock(s)? (If so, is the factory default setting changed before installation?)

52. How many individuals have access to the computer area cipher lock combinations?

53. How often are computer area cipher lock combinations changed? (This should be done at least every 90 days.)

54. Do the doors to the computer area remain closed and locked to help prevent unauthorized entry?

55. To allow system repair and maintenance, is there a key/lock system utilized to protect this access?

56. Who is responsible for, and has access, to these keys?

57. Is entry into the computer area allowed by key access?

58. How many individuals have keys to the computer area?

59. How difficult is it to duplicate computer area keys?

    A. Do keys state "DO NOT DUPLICATE?"

    B. Are they coded keys?

    C. Are they made on special blanks not available to others?

60. Do security personnel help control all perimeter openings to the facility?

61. Are individuals identified by a badging system when in the computer areas?

62. In smaller facilities, is personal recognition utilized?

63. If utilized, is the personal recognition system adequate?

64. Is the computer area entry controlled by human intervention such as guard(s) or other individuals?

65. Who is the individual ultimately responsible for authorizing computer area entry (System Administrator, System Operator, Security Department)?

66. To control access to the computer area, are there strict issuance procedures for keys, cards, badges, and combinations?

67. Who challenges unknown individuals in the computer areas?

68. When leaving the computer area, are all containers, lunch boxes, briefcases, etc., inspected to prevent the unauthorized removal of proprietary information and storage media?

69. Before entering a computer area, are visitors required to sign an entrance log? Are they escorted?

70. Is there documentation to track all after-hour computer area access?

71. When not in use, are all removable recording media such as tapes and diskettes adequately secured to prevent attempted theft?

72. Is access to operations manuals, system specifications documents, user documentation, and system software limited?

73. How is computer-stored information transmitted both inside and outside this facility (coax cable, copper wire, common carrier dedicated lines, common carrier nondedicated lines, fiber-optic, satellite, etc.)?

74. Are satellite components, wiring junctions, modem rooms, PC modems, and other communication link devices located in areas which have limited access?

75. If a shared printer is used, how is access to sensitive output controlled?

76. When sensitive and key financial information is being processed are two people required to be present during all phases of the operation?

77. Are media storage, waste disposal, and paper storage areas segregated from electronic equipment?

78. Is computer system cleaning, maintenance, and repair only conducted by an individual who has been prescreened by security?

79. Is a reliable individual, prescreened by security, used to clean the computer area?

80. Does sensitive information remain secure while cleaning and maintenance functions are being performed? (Explain how.)

81. What is the time of operations for each of the major computer areas? (List on the facility map with locations of computers.)

82. If applicable, are public tours prohibited from accessing sensitive computer operations areas?

83. Are programming personnel generally not permitted to be in the room which houses the computer equipment?

84. Are the duties of personnel working on critical information segmented so that one individual would be unable to control all steps of processing?

85. Are all critical information storage areas prohibited to be used as work areas?

86. Is the handling and destruction of critical information strictly controlled by established procedures?

87. Are critical information processing areas arranged so that a supervisor can monitor all activities at all times?

88. Is a computer area programmer present if any type of software modifications are being made, especially with a vendor?

89. Is a computer area technician present if any type of maintenance or upgrades of equipment are being performed, especially with a vendor?

90. Does the computer area facility have contingency plans for labor strikes and civil unrest?

COMPUTER AREA FIRE SUPPRESSION

91. Does your facility have a written fire prevention/suppression plan?

92. Is there an automatic fire suppression system in all computer areas? (If so, describe.)

93. Are current fire suppression systems adequate?

94. List the computer equipment room's automatic sprinkler head's temperature rating for those located in the ceiling.

95. Is there an extinguishing agent, other than water, used in occupied areas by the automatic system (an environmentally neutral substance, not Halon)?

96. In the event of a large-scale emergency, is there an adequate supply or a close proximity back-up source of the fire suppression agent?

97. To help protect personnel from dangerous extinguishing materials, is highly visible signage and well-known locations used for the storage of breathing apparatuses?

98. Is there a fire training program addressing the use of all fire suppression/detection equipment, systems, and procedures?

99. Does the fire training program address the dangers of gaseous extinguishing agents in occupied areas (including the review, documentation, location, and operation of all automatic sprinkler shutoff valves)?

100. In occupied areas, is there a manual activation mechanism for the automatic fire protection system?

101. In order to suppress fires in open space areas and computer equipment cabinets, has an automatic fire protection system been installed which will cover both of these areas?

102. Where are portable fire extinguishers located within the computer area? (Are they adequate in number, type, and location?)

103. Is there emergency lighting in the computer area to illuminate fire extinguishers in the event of a power outage?

104. How often are portable fire extinguishers tested?

105. Have all personnel been instructed in the use of portable fire extinguishers?

106. Do the computer area's walls and penetrations have a fire rating of at least 4 hours?

107. List the computer area's wall penetration and fire ratings.

108. Are smoke and fire detectors installed in the air-conditioning return ducts?

109. What equipment is available to exhaust smoke and combustion products if a fire were to break out in the computer area?

110. To help prevent water damage in the computer area, are all water (except sprinklers) and overhead steam pipes rerouted?

111. Are wire and pipe penetrations into the computer area watertight?

112. Is there a drainage system in the computer area?

113. Is the drainage system sufficient to prevent water overflow from adjacent areas?

114. Are the computer area floor drains fitted with anti-backflow valves?

115. Is the computer area ceiling constructed to guide water from upper story levels away from all hardware?

116. Are computer area waste containers considered to be low fire-hazard potential?

117. Do these waste containers have metal lids for quick extinguishing situations?

118. To prevent large waste piles, are computer area waste receptacles frequently emptied?

119. Are computer area waste containers emptied outside the computer area to reduce the proliferation of equipment-damaging dust particles?

120. To limit computer area air pollution, are paper-shredding, paper-bursting, and report separation equipment located outside the computer area?

121. Are paper product supplies stored in the computer area? (If so, is this accumulation kept down to a one-day supply?)

122. Are paper supplies stored in the computer area kept in a water-free/water-resistant location (i.e., away from possible sprinkler drainage areas)?

123. Is there an industrial wet/dry vacuum cleaner available in the computer area for use in emergency situations (sprinkler activation, etc.)?

124. If there is an overhead water leakage or discharge, does the computer area have an adequate supply of waterproof covers to protect the equipment?

125. Are there enough waterproof sheets or covers to adequately cover all hardware, equipment, and supplies in the computer area?

126. Is it strictly prohibited to keep any caustic or flammable cleaning fluids in the computer area?

127. To reduce the threat of fire, are caustic and/or flammable cleaning fluids limited to only one day's need?

128. Are these agents stored in approved containers?

129. Is there a certified-fireproof, magnetic-media cabinet/safe located in the computer area for the storage of critical documents, disks, and tapes?

130. Are employees familiar with fire suppression techniques common to computer media?

131. Are firefighting water sources strategically located for computer areas?

132. Is the community fire department located farther than three miles from the computer facility?

A. Where?

B. What is the response time?

C. Are they staffed to appropriately handle your facility?

133. Is the fire department staffed with full-time or volunteer personnel?

134. How often are computer area fire drills practiced?

135. Have individuals been assigned specific responsibilities in case of fire (i.e., fire brigade leaders, evacuation leaders, etc.)?

136. Among the computer area's personnel, are there individuals with emergency medical assistance, cardiopulmonary resuscitation (CPR), and/or first-aid training?

137. Is the staffing scheduled so that at least one of these trained individuals is on duty at all times?

138. Are first-aid supplies located in a manner that allows quick access in medical emergencies?

139. Is there a computer area evacuation plan?

A. How often is it practiced?

B. Is it in writing?

140. Are there fire doors to compartmentalize the building in case of fire or flooding?

141. How are these doors closed (automatically or manually)?

142. Is a complete set of vendor-recommended spare parts available to maintain computer area emergency systems (CCTV, alarm, fire suppression, etc.)?

COMPUTER EQUIPMENT SECURITY

143. How are computers and computer components controlled? (See Inventory Control.)

144. Who is assigned to select, purchase, and control computers and peripheral components?

145. Does this same individual specify and write bid specifications for this equipment?

146. Is computer equipment controlled by a department other than the computer department?

147. Are end users of computer assets given training on possible security vulnerabilities?

148. Has an inventory been conducted to include all company-owned computer assets?

149. When was the last computer asset inventory performed?

150. Who performed the last computer asset inventory?

151. How often is the computer asset inventory performed?

152. Were any irregularities identified since the last computer asset inventory?

153. Are there separate categories for the facility's mainframe, network system, and personal computers within the inventory system?

154. What is the location of each of these systems?

155. Is all supporting hardware and software considered when determining each category's total value?

156. Are computer-associated equipment and devices listed in their own separate valuation category?

157. Is a unique identifying number permanently affixed, stamped, or etched to each piece of equipment for inventory tracking purposes?

158. Are personal computers able to be physically secured by locking devices?

159. Are antistatic pads, mats, or carpets used in personal computer areas?

160. Is the company actively trying to limit personal computer usage to a single user?

161. Are key locks on PCs utilized when they are not in use?

162. Is there a lock on the power switch prohibiting unauthorized equipment use?

163. Does the type of information processing performed in the computer area require an uninterruptible power supply?

164. If so, what type of an uninterruptible power supply system does the computer area utilize?

165. What type of power filtering devices does the computer area utilize to protect against power abnormalities?

166. Does the computer area use power-line filters?

A. If so, what type?

B. Purchased when?

C. Who is the vendor?

167. Does the computer area use isolation transformers?

A. If so, what type?

B. Purchased when?

C. Who is the vendor?

168. Does the computer area use constant-voltage transformers?

A. If so, what type?

B. Purchased when?

C. Who is the vendor?

169. Does the computer area use motor-driven generators?

A. If so, what type?

B. Purchased when?

C. Who is the vendor?

170. Where are emergency power-off switches located in the computer area?

171. How are they protected from accidental activation?

172. Is the local power supply adequate, consistent, and reliable for use in your facility? (Who made this observation? What information was used in making it?)

173. Have special procedures been implemented to protect the following electrical equipment:

A. Transformers?

B. Motor generators?

C. Breaker panels?

D. Cooling towers?

174. Before installing computer equipment, has the structural flooring in the computer room been deemed to be of adequate strength to support the loads imposed by the various pieces of equipment?

175. Does the landscape around the exterior of the building adequately remove water accumulation during sudden or seasonal heavy rainfall?

176. Where are facility storm drains located (include in the facility map)?

COMPUTER INTRUSIONS AND THEFTS

177. Are employees instructed to contact specially trained personnel if security is breached or compromised?

178. Do personnel know the proper procedures for handling a computer system penetration in progress to track and minimize damage?

179. Does the detection of any of the following computer intrusion or theft activities require the immediate notification of security?

 A. Prohibited use of computer access time resources

 B. Illegal use or attempts to access information files in the computer system

 C. Misuse of computer-related supply resources

 D. Misuse of computer software intended for company use only

 E. Misuse of computer hardware resources

 F. The unauthorized destruction of software or information resources

 G. The input of unauthorized information into the company's computer system

 H. Altering any of the company's information resources

 I. Inputting unauthorized files and records into the company's computer

 J. Examining waste receptacles for company sensitive information

180. Does the employee computer security briefing address each of these areas so that they will be able to recognize any of these activities?

181. Are media storage inventories controlled, maintained, and audited by a system designed to be accurate, updateable, and secure?

182. Are completed labels applied to all magnetic storage media for proper identification?

183. Are there established procedures so that only authorized individuals are responsible for the storage media's marking, handling, storing, destroying, and transmission?

184. What are the procedures utilized?

185. Is storage media locked when not in use?

186. Are tape and diskette locations and storage areas documented in a continuously updated filing system?

187. Is shredding, or magnetic erasure (degaussing), the proper method of destroying disks containing confidential company information?

188. Are all printouts, magnetic media, and printer ribbons with sensitive information destroyed by using approved security methods?

189. How is authorization for system access granted?

190. How is it limited within this facility?

191. Who has system access? (Why these individuals?)

192. Who grants system access?

193. Who is responsible for computer resource use and supervision?

194. Is there an established written sign-in log tracking who has accessed company computers?

195. Can users who have access to applications programs receive direct financial gain (inventory, products, negotiable items, or cash) without approval from management simply by using the applications?

196. Could an employee once granted access to the computing resources stand to benefit in terms of actual promotion or perceived employment status changes if they were to misappropriate those resources?

197. Within the past five years, has any computing resource fallen under "attack" to an outside source (hacker)?

198. Have you contacted the local police department to determine if local hackers are a threat? (Have they been active? Are there any ongoing investigations? Is white-collar crime a major concern in the local area?)

199. Can unauthorized access to or unavailability of sensitive information result in a perceived triumph for legal action groups or individuals outside of your company?

200. If your systems were compromised, could it prove potentially embarrassing or legally damaging?

201. Within the past five years, has either an employee or an outsider attempted or been successful in damaging or destroying data, the computing area, or the facility?

202. Is your facility located in or near a college community? (These areas experience more hacking incidents.)

DISASTER RECOVERY PLANNING

203. Does the company have a computer disaster recovery program?

204. Who has been assigned to develop the computer disaster program?

205. Does the computer disaster program take into account the following types of major disasters?

    A. Aircraft transportation accidents impacting facility

    B. Chemical splashes or contact with toxic materials

    C. Deficiency of public utility to adequately supply services

    D. Dust

    E. Earthquake

    F. Falling objects and flying particles

    G. Fire

    H. Electrical hazards

    I. Hazardous or explosive materials

    J. Hurricane

    K. Landslide

    L. Natural disaster flooding

M. Severe freeze prevention procedures

N. Snow storm

O. Unknown atmospheres that may contain inadequate oxygen to sustain life or toxic gases, vapors, or mists

P. Water flow from a sprinkler system

206. Does the computer disaster program take into account the following types of criminal activities?

A. Arson

B. Blackmail

C. Bootlegging

D. Breaking and entering

E. Bribery

F. Burglary

G. Collusion

H. Conspiracy

I. Disorderly conduct

J. Embezzlement

K. Extortion

L. Fraud

M. Grand larceny

N. Larceny

O. Libel

P. Looting

Q. Malicious mischief

R. Pilferage

S. Piracy

T. Racketeering

U. Terrorism

V. Theft

W. Trespassing

X. Vandalism

Y. White-collar crimes

207. Has a financial impact estimate been established to address each of the above activities?

208. Does the disaster recovery program address Wide Area Networks (WANs), Local Area Networks (LANs), as well as microcomputer systems and their vulnerability to disasters?

209. Has a disaster recovery "test" been conducted to identify the effect of a possible operation cessation within each of the company's computer systems?

210. Is there a contingency plan to replace critical hardware in the event of a large-scale computer system disaster?

211. Does the company maintain inventories of critical hardware on-site to replace damaged components?

212. If not, how will the company obtain this equipment?

213. As part of the computer disaster recovery plan, has the effect of both short- and long-term computer interruption been analyzed?

214. How significantly would the company be impacted if each computer systems was terminated for:

A. One day?

B. Two days?

C. One week?

D. One month?

215. In a case of civil unrest or natural disaster, are critical employees furnished with company letters requesting allowance to transverse geographically controlled areas to allow them to reach the company facility?

216. Who is responsible, in natural or manmade disasters, for instituting start-up procedures?

217. How far away from the facility does this individual live?

218. Are back-up individuals assigned to this function if a disaster occurs and if the "primary" is unable to perform their duties?

219. How will these individuals make it to work should the roads be closed?

220. In a disaster situation, the following individuals may need to respond to the facility (depending on facility size):

A. Administration support

B. Computer analysts

C. Computer operators

D. Computer programmers

E. Computer users

F. Customers of the facility (depending on the facility's function)

G. Facility vendors/suppliers

H. Maintenance staff

I. Management

J. Public utility personnel

K. Security officers/administration

221. How will they reach the facility in a major disaster?

222. Is there a reporting structure/hierarchy for these individuals?

223. What is the plan for their access to the facility?

224. In the event of a major disaster, has a disaster response control center been established as an operational center?

225. Who has the authority to open this area?

226. Who will staff this facility?

227. Have back-ups been assigned in the event that "primaries" are unable to respond?

228. How will communication be accomplished?

229. How many telephones are located in the operational center?

230. How many cellular phones?

231. How many radios?

232. Can the radios on a repeater system be disabled in a disaster?

233. Is there a current telephone book maintained in this area?

234. Is there a current contact list maintained in this area?

235. Have individuals been assigned as "runners" to perform errands and to communicate information?

236. Is all equipment and information positioned in convenient locations?

237. Is all sensitive information (home phone numbers, call-out lists) maintained in secure locations until this area is put into operation?

238. Have management personnel been trained to restart and operate the critical computer systems if computer staff is not normally maintained 24 hours a day?

239. Are computer personnel crosstrained in emergency procedures and system operations?

240. Has a visitor's log been set up in key areas to record entries and exits during a disaster?

241. In a disaster situation, have the following procedures been established for all computer programs and key computer-related personnel?

A. Egress facilitators (badges, key cards, keys) for normally off-limit areas

B. Access control to ensure that noncritical personnel are barred from key areas

C. A visitors ingress/egress log system

D. A checkout system for all computer documentation

242. Have employees completed training in the following areas?

A. All emergency procedures

B. Emergency security procedures

C. Facility equipment shut-down procedures

D. The location of all emergency-related equipment

E. The use of all fire and emergency equipment

F. The use of communication equipment for emergency situations

243. Is the loss of hardware, peripheral equipment, and software program's replacement costs sufficiently covered by the current insurance policy?

244. In the current insurance policy, what is the deductible amount for the loss of computer equipment?

245. Who is the designated Disaster Recovery Manager (DRM) responsible for the coordination and implementation of the disaster recovery plan?

246. Are natural disasters and severe weather such as frigid temperatures, blizzards, floods, tornadoes, and hurricanes a threat to the region where the major computer resources are located?

247. If so, have any of these weather conditions rendered the computing area or computing resources inoperable for more than five days at a time in the last five years?

248. Is there an active earthquake fault, high erosion area, or active volcano within 50 miles of the company's major computing resources?

249. In relation to the computer area, are there any dams, oceans, lakes, or bodies of water that could cause flooding?

250. Is the computer area within 2,000 feet of a railroad or roadway in which volatile chemicals, liquefied natural gas, or explosives are transported?

251. How close is the closest major international, commercial, or military airfield?

252. Are there defense contractors or installations, government laboratories, or nuclear facilities located within a ten-mile radius of the computer area locale?

253. Has a Computer Area Emergency Contingency Plan been prepared?

254. How does the Computer Area Emergency Contingency Plan approach the following?

A. The reporting of emergencies and the notification to key personnel

B. Emergency voice and data communications requirements and backup procedures

C. Computer area evacuation

D. Fire emergencies

E. Water/flood emergencies

F. Power failures

G. HVAC failures

H. Structural instability or damage, such as that caused by earthquakes

I. Weather or other natural phenomena caused emergencies

255. Is there an annual review by computer resources personnel of all emergency-response procedures?

256. Are emergency plans “drills or simulations” involving all personnel and resources performed annually?

257. How often are practice drills conducted for the following events?

A. Catastrophic incidents

B. Bomb-related threats

C. Civil disturbances

D. Terrorist actions

258. In the computer area, is a list of emergency phone numbers and location identifiers posted for each of the following?

A. Medical aid

B. Nearest fire response personnel or fire department

C. Police

D. Facility security personnel

E. Flood

259. How frequently is the disaster recovery plan tested for effectiveness?

260. Who performs and supervises the testing of the disaster recovery plans and making any changes in these plans?

261. Is this information stored in an off-site, fireproof area in case that a disaster occurs at the main facility?

262. If disaster recovery service is used, how far is it from the facility's location?

263. If utilized, what type of disaster recovery service is used (commercial vendor, service bureaus, shared contingency agreements with other companies)?

264. What type of special needs does this particular facility have during one of the named emergency situations (supplies, equipment, etc.)?

265. Has a contact list of management/owners been created so they can be immediately notified in all emergency situations?

266. In order to keep emergency contact lists updated and current, what procedures have been implemented, and who is primarily responsible for maintaining them?

267. Are all critical business operation files kept in either original form, hard copy, or backup form?

268. Are all critical files controlled by an Electronic Asset Protection (EAP) system when they are transferred to floppy disks to ensure that they are not surreptitiously removed from the facility?

INFORMATION INPUT CONTROLS

269. Have all employees been instructed on the proper procedures for information input?

270. Is the system configured to identify information input mistakes?

271. Does the system and/or computer area personnel help the user correct any mistakes?

272. Is inputted information integrity continuously checked by an information validation program?

273. Are certain types of information input only allowed from certain terminals?

274. Are certain types of information input only allowed from certain users?

275. Is there a formalized method of reporting information input problems for users?

276. When there are system problems, how are information input users notified?

277. Is system hardware consistently monitored for acceptable operation?

278. What preventative maintenance is performed on the system's input terminals?

279. Are hardware components durably constructed?

280. Do terminals utilize a screen-saving feature after a designated period of inactivity?

281. Are screen savers password protected?

282. Is the system configured so that passwords cannot be printed or displayed on terminals?

283. Does the system deny access after three unsuccessful log-on attempts from any terminal?

284. Are these unsuccessful attempts recorded and investigated by a system administrator?

285. Is each individual's user identification automatically tracked to all the work they perform on the system?

286. Does the system track the total usage of each individual?

287. When a user is processing their inputted work, is a backup made of all information before it is processed (for a report) in case of a system problem?

288. What are the procedures for user input information recovery?

289. If the main system does fail or is shut down for an extended period of time, can personal or microcomputers be used?

290. Do remote sites have limited access based on their specific processing needs?

291. Are remote site's equipment physically secure from unauthorized access?

292. If the equipment is not located in a secured area, are special physical locking devices used to prevent access?

293. Whenever possible, are terminals and remote sites locked out of the main system after the designated operating hours?

294. Is input information encrypted if its source of origination is a remote site?

295. Does the system check all information input for viruses from both on-site and remote facilities?

296. Are all users correctly trained on how to properly access their system applications?

297. Can the system be easily changed to meet input user's requests?

298. Can only authorized computer area personnel make any system changes or input terminal configurations?

299. Does input information have to meet specific system requirements?

INFORMATION PROTECTION

300. Has the company established a proprietary information protection program?

301. Has all proprietary information been clearly identified?

302. Have all of the possible sources of this information been identified and safeguarded?

303. If the facility maintains compartmented or special company sensitive information, list restrictions on the use of such information.

304. Is the protection of competitor sensitive and proprietary information sufficiently addressed by the company's accountability procedures and tracking system?

305. Did all employees sign a confidentiality agreement when first hired?

306. Has this agreement been carefully reviewed and approved by legal council?

307. Is it mandatory for the confidentiality agreement to be re-reviewed and signed when an employee leaves the company for any reason?

308. Are periodic meetings given or memorandums sent out to all employees to remind them of their responsibilities to protect proprietary information?

309. Does the employee security briefing explain safeguarding protocols as they apply to the protection of competitor sensitive and proprietary information?

310. Have all closed, limited access, and secured areas been visually designated by the use of prominent signage?

311. In secured areas, what procedures are used to control visitor admission and registration?

312. Is there a system of logging or recording visitor parking passes issued to visitors?

313. Are company personnel and visitor entrance logs saved for reference purposes?

314. List the type of alarm systems that these areas are currently using.

315. Are any unusual security vulnerabilities evident in regards to company-sensitive information?

316. When information is transported outside the facility, what method is used to protect proprietary or competitor-sensitive information?

317. Is information transmitted via a designated courier system, which has been previously screened and approved by security?

318. Describe the courier systems used for compartmented or special company-sensitive information.

319. Have the marking caveats of "secret, proprietary, company sensitive, or limited" been used to help protect confidentiality of company documents?

320. List the specific caveats used to indicate the level of sensitivity for the different types of information.

321. Is all proprietary information clearly marked in red ink on the document or electronic medium?

322. Are individuals from outside the company (consultants, vendors, temporary labor) held to the same standard of information protection is expected from employees?

323. Is this provision clearly outlined in all contracts and labor agreements?

324. Are there specific procedures for the transmission of any proprietary information to other company locations or to other businesses?

325. Do these procedures pay special attention to electronic forms of transmission?

326. Are there special procedures for protecting information during required inspections from governmental entities?

327. Is the company involved in military or energy-related projects which requires adherence to government-established information protection programs?

328. How is sensitive information protected during its dissemination?

329. Have policies been implemented to prevent the unauthorized release of drawings and other technical proprietary data during trade conferences and outside meetings?

330. Who is responsible for implementation and supervision of the proprietary information protection program?

331. Are all employees' offices, desks, and file cabinets kept locked at all times when not in use?

332. Is information not left laying around and visible to casual observers?

333. Is mail removed from open boxes or trays?

334. Are all planning charts and writing boards destroyed or erased immediately after their use?

335. Have the proper caveat markings been applied to all photographic material?

336. To protect small, easily lost sensitive information items, what safeguards are used for their dissemination?

337. How is the storage area for sensitive information secured? Where is it located?

338. When the company's proprietary information is lost, compromised, or disclosed, what methods are utilized for this particular situation?

339. Are all incidents of compromises involving proprietary information immediately investigated and documented thoroughly by the company?

340. How is this type of incident investigated?

341. Who investigates this type of incident?

342. Once an incident has occurred, what type of effort is made to prevent any further compromise of information?

343. If potential criminal actions are involved in a compromise, is the appropriate law enforcement agency notified immediately?

344. Is the secure disposal of confidential information thoroughly addressed by the company's policies and procedures?

345. Who reviews and collects all the company-sensitive information waste?

346. Is all proprietary information completely destroyed before it is disposed?

347. Before destruction, are special containers used to securely store all proprietary information?

348. Is access to sensitive information waste strictly controlled?

349. What method is used for collecting sensitive information waste in the facility?

350. Is proprietary information destruction constantly supervised?

351. Who is responsible for performing the destruction of sensitive information waste?

352. Is shredding used to destroy sensitive information waste?

353. Is the destruction of all sensitive information waste recorded in a log book and/or file?

354. During an emergency, what procedures are used for special sensitive information waste destruction?

355. Where are all of the sensitive information waste emergency destruction areas located within the facility?

INFORMATION SYSTEM ACCESS

356. Has the company performed a computer vulnerability assessment?

357. When was the last computer vulnerability assessment performed?

358. How often are these assessments performed?

359. Were all systems, including each microcomputer system, included in the vulnerability assessment?

360. Who was responsible for conducting the computer vulnerability assessment?

361. When conducting the computer vulnerability assessment, were the following hardware vulnerabilities addressed?

   A. Central Processing Unit (CPU) failure

   B. Memory failure

   C. Storage device failure

   D. Partial storage device head crash

   E. Storage medial transportation or storage disaster

   F. Circuit failure

   G. System vulnerabilities to natural disaster

   H. Power surges or loss of power and possible associated damage

362. Did the computer vulnerability assessment address possible damage to major system software and application programs? These may include:

A. System defects

B. Logic errors

C. Poor program design

D. Logic bombs, viruses, and trap doors

E. Incomplete program documentation

F. Loss of documentation

G. Errors in algorithmic operations (including truncation and/or rounding)

363. When covering critical systems, were the following vulnerabilities addressed?

A. Software interoperability

B. Operator or information entry error

C. Critical file erasure

D. Intentional sabotage

E. Hardware malfunction

F. Physical damage to magnetic storage media

G. Natural disaster

H. Overwriting of critical files

I. Operator misuse of equipment or programs

364. Did the computer vulnerability assessment address telecommunications equipment, which include any security vulnerabilities in the following areas?

A. Power spikes, black-outs, brown-outs, and transient surges

B. Noncompatible telecommunication equipment and/or communication protocols

C. Inappropriate switch settings

D. Improper maintenance

E. Hardware and software problems

F. Transmission line noise problems

G. Transmission line faults

H. Eavesdropping, the surreptitious interception or diversion of critical information, and/or system wiretapping

365. When in unsecured areas, are special precautions implemented to protect sensitive materials that are printed and displayed on terminals from visual access?

366. In those systems containing critical or sensitive information, is there an absolute division of duties (i.e., auditing, information entry, programming, etc.)?

367. Are individuals working with critical or sensitive information bonded?

368. Is critical or sensitive information removed from desk tops, white boards, and flip charts during times when it is not essential that this information be displayed?

369. How is the copying of files and programs controlled?

370. Is there an audit trail that records any program changes?

371. How is consent authority obtained to change critical program or information files?

372. Who can give this consent?

373. Who has access to critical program and information files?

374. Who reviews these changes?

375. Who tests changes to ensure that they do not adversely affect program operations?

376. Because testing should only be performed on backup data, what system has been implemented to ensure that it is not performed on active data?

377. During the computer vulnerability assessment, were the following operator/end user vulnerabilities addressed?

A. Misplaced storage media

B. Mislabeled storage media

C. Negligent equipment care

D. Information entry errors

E. Malicious handling of information and equipment

F. Programming errors

G. Improper or nonadherence to procedures

H. Poor housekeeping issues (spilling liquids on equipment or eating around equipment)

I. Malicious attacks by disgruntled employees

J. Intentional criminal activities

K. File deletion caused by operator inattention

L. File write-over caused by operator inattention

378. When computer systems convey, hold, or process critical company information (financial, personnel records, or strategic), what constraints are put on their operation?

379. Who has access to these systems?

380. Do these systems allow access by a dial-up modem or outside phone line?

381. Are the system and physical security controls adequate to protect these critical records?

382. How often are critical records backed-up?

383. Does your company have independent auditors (other than end users) who inspect the internal operations of computer systems?

384. Are independent, noncompany, auditors utilized to verify internal audit reports?

385. Do auditors also audit all stand-alone and microcomputer systems?

386. If company computers are used by noncompany staff, who ensures that security procedures are followed?

387. Who audits these systems?

388. How is work controlled when it is performed by noncompany staff?

389. What type of mainframe equipment is utilized at the facility?

390. Who is the system operator for the mainframe?

391. Have there been incidents of repetitive (10 or more) unsuccessful logon attempts to a single account, user's reporting account activity between logon sessions (normally off hours), or other possible security breeches to the system operator?

392. What are the controls governing job entry for batch, remote-entry, and on-line processing?

393. Has a protection scheme been implemented for the information in the system tables? (If so, what is it?)

394. Are all computer area personnel trained in the methods used to circumvent hardware and software security functions in the systems used in the computer area? (Can they recognize these attempts? Do they look for them?)

395. Is the hardware used in the computer area known to have "trap doors" for evading operational system security mechanisms that are generally or publicly known? (e.g., Do they appear on hacker bulletin boards?)

396. Are all hardware malfunctions documented and recorded in a tracking log?

397. How many individuals residing in your area (both employees and other persons) are knowledgeable enough to use these system vulnerabilities (estimate for risk analysis)?

398. Is the mainframe operating system stored in read-only memory?

399. How often are checks of operating system integrity made by the system managers?

400. What are the normal defensive actions taken for operating system violations?

401. Does the operating system automatically record and suspend program activity when attempts to access unauthorized files are identified?

402. What defensive action does the operating system take in response to unauthorized security-table access attempts?

403. Is there a system warning message generated on all terminals for violations?

404. Does the system automatically log-off idle users (this should be set to 15 minutes)?

405. Have all original vendor and test accounts been disabled to preclude use by unauthorized personnel?

406. Are all passwords assigned to individual users rather than group or "department" accounts?

407. What is the process to delete "short-term" accounts (temporary employees, summer interns, etc.)?

408. Is there only one designated security terminal to change system security perimeters and functions?

409. Is this security terminal operated by a member of the security department or other individuals who are not system programmers?

410. Can the system's internal clock be reset from any terminal?

411. When users are signing-off, does this procedure prevent anyone from being able to sign-on by manipulating a cancellation message?

412. Are all critical information programming standards thoroughly examined by security personnel responsible for computer area operations?

413. Are new employee's activities carefully monitored, especially corrections and errors?

414. Are preprocessed and postprocessed information input figures compared for discrepancies?

415. Are programmers prevented from having any control over any information which may affect them personally, such as bonuses or benefits?

416. Are any unusual types or amounts of system activities examined by an independent source?

417. Are logs of all system activity maintained and reviewed by management?

418. Are these logs utilized to statistically evaluate system processing performance?

419. Do these logs serve as the basis to expense system costs to the applicable user groups by level of usage?

420. Are these logs secured to prevent any changes, damages, or losses?

421. Are all unauthorized changes in, or damage to, the logs immediately investigated?

422. Is there a formal employee termination checklist which must be completed before the employee receives their last paycheck or departing benefits? This checklist should include:

A. Company-issued identification and verification items

1. Have all company identification (badges, etc.) been returned?

2. Have all parking or vehicle passes been returned?

B. Access control items

1. Have all access cards been returned?

2. Have all keys allowing access to company property been returned?

3. Do cipher lock combinations need to be changed?

C. Electronic access control items

1. Have all voice mail passwords been changed or deleted?
2. Has the user's voice mailbox been deleted or access code changed?
3. Have all high-level PBX system passwords been changed or deleted?
4. Have all PBX system identification codes been changed or deleted?
5. Have all PBX access codes been deleted?
6. Have all computer identification codes been changed?
7. Have all computer accounts been deleted?
8. Have all computer accounts available to the employee been reviewed to ensure they have not been deleted?
9. Has all the computer equipment been returned?
10. Has all of the company-owned software been accounted for?
11. Are all company-sensitive documents accounted for?
12. Has all of the company-owned hardware been accounted for?

D. Other company-related items and issues

1. Have all business-related memberships been canceled?
2. Have all outstanding business-related bills been reconciled?

3. Have all business-related credit cards been returned? These should include:

    a. Visa?

    b. MasterCard?

    c. Diners Card?

    d. American Express?

    e. Other?

E. Company benefits, concessions, services provided by the company for employees only

1. Have all long-distance calling cards been returned?

2. Have all company-related subscriptions been canceled?

NETWORK SYSTEMS (LOCAL AREA NETWORKS)

423. Who controls access to the network? (Who is the system operator?)

424. Is the network password-protected, or is there "open access" to the node and/or net?

425. Does any LAN wiring run outside secured areas (facility walls)?

426. Are LAN users identified?

427. Are LAN applications and supporting files identified?

428. How much of the company system's information will be shared on the LAN?

429. Is there a separate LAN setup for critical data?

430. Is critical information transmission accomplished via a "router?"

431. Is LAN access determined by the need of an employee's job duties?

432. Is removable media utilized to store critical information used on the LAN?

433. Is LAN access restricted by the use of a menu-driven applications?

434. Is downloading from the LAN restricted by access to a limited number of locally secured drives?

435. Have the appropriate control procedures been implemented to protect the LAN file server?

436. Is a removable device for LAN backup storage protected when it is not being used?

437. Is complete LAN access only given to the network operations manager and his/her assistant?

438. Are all employees given a briefing on the significance of LAN security concerns and procedures?

439. Are strictly controlled password procedures utilized for LAN access?

440. Are LAN passwords closely monitored for unusual activities or patterns?

441. Are any modems which have access to the LAN (including those connected to PCs) disconnected when not in use?

442. Are LAN security features and devices incapable of being neutralized or circumvented without detection?

443. Do all security features remain operational even if the power is turned off and then back on again?

444. Is there an alarm on the LAN that will be triggered if unauthorized access is attempted?

445. Does the LAN record all unauthorized activity, especially hacking?

446. Once a hacking incident has occurred, is it thoroughly investigated to determine who is the source of this activity?

447. Is the hacking source determined to be inside or outside the company?

448. Did the hacking originate from a vendor who may be somewhat familiar with the company's operations?

449. Could the hacker possibly be a disgruntled and/or former employee?

450. Is senior management and law enforcement immediately notified when a hacking incident occurs?

451. Are countermeasures immediately implemented to prevent any further intrusions to the LAN?

452. If a computer security expert is not on staff, has the company contacted a reputable consultant to assist in reducing LAN vulnerabilities?

PASSWORD PROTECTION FOR INFORMATION SYSTEMS

453. Is password file access adequately controlled to prevent its compromise?

454. List all the individuals with password file access.

455. Is an encryption scheme used to protect the password file?

456. To protect their integrity, are passwords prevented from being displayed on any screens?

457. Has a password protection policy been implemented?

458. Is password misuse defined in writing for employees, along with the subsequent disciplinary measures to be taken for any violations of this policy?

459. Is there an automatic time-out or user reauthentication system implemented when there is a specific period of no terminal activity?

460. Are computer system users required to sign a receipt of password issuance?

461. Are all computer account password requests submitted in writing and verified for authenticity?

462. For each computer system is there an audit trail for all activity?

463. Who is responsible for maintaining and reviewing the integrity of audit trails and written computer access logs?

464. Are different levels of system access (user, super user, system operator/SysOp) maintained by an access level control program?

465. Do password protection configurations use personal identification numbers (PIN) to identify and grant access authority to the various levels of system users?

466. Before system users leave their terminal area, are they required to log-off their terminal's access to the system?

467. When system users are transferred, quit, or are terminated, are their passwords and other system access capabilities suspended or completely disabled?

468. Are passwords generated by the user or are they computer generated?

469. Have all individual user passwords been changed once every 30 days?

470. Does the system automatically prompt users to change their passwords after 30 days?

471. Is the use of (at least) six-character, uncommon passwords required?

472. Does the system "help" users select passwords that are not easily guessed or account repeats (many users will try to use the same password repeatedly)?

PHYSICAL SECURITY FOR COMPUTER INFORMATION SYSTEMS

473. Does the security plan include special considerations for computer operations?

474. Are computer areas considered to be critical and restricted areas?

475. How many of these critical and/or restricted areas are there, where are they located (attach drawings or plans)?

476. List all critical area's restrictions, barriers, and controls such as locking devices, alarm systems, and guard patrols.

477. What method of designation is used to identify these areas?

478. How is control of these areas properly administered?

479. To prevent access by windows, skylights, and ventilation shafts, what safeguards have been implemented?

480. Are these areas circumscribed by any perimeter barriers?

481. Are these areas adequately protected by intrusion alarms?

482. Does the facility have compartmentalized computer areas which are physically separated from other less secure areas?

483. Are these areas marked as “Restricted Areas” or “Controlled Areas?” (How is access controlled? Are these areas separately alarmed?)

484. Is the computer area protected by an intrusion alarm system, which is monitored by a central station?

485. List the restricted area’s type of alarm system and any related components used.

486. What characteristics or special attributes are used in the construction of storage areas?

487. Is there an electronic access control system?

488. Does the access control system use keypads, card readers, and biometric devices?

489. What type of access card technology is utilized for the access control system (magnetic stripe, smart card, Wiegand, proximity, etc.)?

490. Does the access control system track personnel by individual personal identification number (PIN)?

491. Does the access control system record and create an audit report of all ingress and egress activities by PIN, date, time, and access point?

492. Is all user access carefully controlled and monitored, especially when hiring and terminating personnel?

493. What are the hours of operation of the computer department?

494. If the computer department operates after regular business hours, are additional security procedures implemented?

495. Are guards assigned to monitor these areas periodically (how often and what is the means of ensuring that the guard patrols have been completed, etc.)? (See also Security Officers.)

496. Are these patrols limited to the perimeter of the computer area?

497. If security patrols the interior of the computer area, are the patrol officers required to use a time/place recording device, such as a Detex clock, to record and track their activities?

498. Are all generated reports and related input data safeguarded before, during, and after they have been processed?

499. Are all information and reports only released to authorized personnel?

500. Are the information processing, equipment room, and data storage areas separated from each other?

501. Within the computer department, are all computers in a separate room with special access required for an extremely limited number of personnel?

502. Does the computer processing equipment room and/or the equipment itself have electronic field emanation reduction features (Tempest) to prevent easy electronic eavesdropping of data-processing activities?

503. Is the entire computer department periodically checked for electronic eavesdropping devices?

504. Are all computer main and back-up power supplies located in a highly secured area and routinely checked for any signs of tampering or sabotage?

505. Are all phone lines, cellular and wireless equipment, and phone switching systems (PBX) located in highly secured area and routinely checked for any signs of tampering or sabotage?

506. Are phone lines periodically inspected for any wiretaps or other eavesdropping activity?

507. Are all satellite equipment and other telecommunications systems located in a highly secured area and routinely checked for any signs of tampering or sabotage?

508. Is there a complete inventory of all computer and telecommunications equipment, which includes a description, serial number, and its location?

509. Is this inventory kept by security and changed every time a piece of equipment is acquired or liquidated?

510. Are all pieces of equipment securely fastened or locked down to a structural component of the building where it is located?

511. Have contingency plans and arrangements been made for immediate replacement for critical computer operating system components if they should be rendered inoperative?

512. Is security, not the computer department, responsible for the distribution and functional operation of all locks, keys, and access control systems?

PHYSICAL STRUCTURE SECURITY FOR INFORMATION SYSTEMS

513. How many stories high is the building in which the computer area is located?

514. How many of the floors are located above the ground level of the building that houses the computer area?

515. How many of the floors are located below the ground level of the building that houses the computer area?

516. List the building floor location of the computer area?

517. If the computer area is in a multistoried structure is it located on an upper level to minimize exposure to floods and intrusion?

518. Is the building constructed on a solid foundation? (Describe.)

519. What type of material is used in the walls of the building housing the computer area (concrete block, reinforced concrete, brick, or stone)?

520. What is the material used for exterior wall construction of the computer area (concrete block, reinforced concrete, brick, stone, wood, or stucco)?

521. Are all building entry doors and/or gates constructed of a metal or metal-clad material?

522. Are exterior building doors constructed of solid wood, glass, or metal?

523. What is the principal material of the interior walls of the building (reinforced concrete, concrete block, brick, or metal)?

524. What is the material used for the interior side of the walls of the building (sheet rock, plaster, veneer on plywood, ceramic tile, etc.)?

525. What is the principal material of the building's ceilings and floors (reinforced concrete or metal)?

526. If not located in a permanent structure, is the computer area housed in a semipermanent transportable building, trailer, etc.?

527. Are windows leading to computer areas accessible from the outside?

528. To prevent confidential information from being observed by an external source, are sensitive area windows made of a translucent material?

529. Are windows leading to computer areas alarmed?

530. Are these windows mesh-embedded glass or reinforced with durable external screens?

531. Are all remote-site terminals physically secure?

532. Do these remote sites have a physically secure library for information and documentation?

533. Are computer terminals and peripherals secured to the building's structure to prohibit or delay theft?

534. Have all references to the computer area's location been removed from any facility directories, signs, or maps?

REMOTE SYSTEM ACCESS SECURITY

535. Does your system have dial-up capabilities?

536. How many incoming ports (lines) are available?

537. What types of systems are remotely accessible?

538. What type of telecommunications equipment can access these systems?

539. Does the operating system have a user-authentication process?

540. Does it require a user identification name or number to grant access to the system?

541. Is information transmitted over the regular telephone lines encrypted?

542. Is a port-protection device or methodology used in the host computer?

543. Does the port-protection device prevent system access from an auto-dial modem or "War Dialer?"

544. Does the port-protection device utilize a call-back feature (call-back modem) to verify the identity and telephone number of the user before authorizing system access?

545. Are dial-up numbers changed periodically?

546. Is the system dial-up identifier devoid of the following?

A. The business name or logo

B. Any indication of any system capabilities or software being used

C. Any information that would help an unauthorized person gain system access

547. Is routine monitoring of dial-up modem access activity to help identify any intrusion attempts and/or successes from unauthorized personnel such as hackers?

548. Does the system software automatically disconnect a dial-up connection after three unsuccessful sign-on attempts?

549. Can a system user's access only be reinstated by a system operator after several unsuccessful logon attempts have caused the disabling of their access pathway?

550. To identify and locate the source of repeated unsuccessful sign-on attempts, is there an audit trail generated to aid investigative efforts?

551. Is the proper protection level caveat automatically assigned to all sensitive information when it is generated by the computer system?

552. Are the proper caveats automatically generated on the top and bottom of each page?

553. Does the cover sheet for sensitive material display the name and address of the facility responsible for its generation?

554. Does an individual have to demonstrate a "need to know" before they are granted access to sensitive computer-generated material?

555. Is a commercially compatible encrypted facsimile unit used when sensitive material is faxed out of the facility?

556. What are the methods used to protect unattended, automatically generated, sensitive material such as off-hour generated reports and faxes?

557. If information must be transmitted via unsecured telephone lines, is it required to be in an encrypted form?

558. What communications software is utilized to transmit information?

559. Is there any identified vulnerability risks posed by the use of this communications software, especially if the software is commercially purchased?

560. Is computer equipment shielded to minimize the ability of electronic eavesdroppers from being able to detect and "read" computer system electrical signal emanations (Tempest security)?

561. For those systems that allow dial-up access to critical systems, does the system report the time of the last log-in?

562. Do the telecommunication transmission (telephone) lines to the company computer system pass through a switchboard?

563. Can the system operator manually disable an individual communication line?

564. How are users certain that they are connected to the correct computer and not to a "hostile" system?

565. Is there an error-detection and error-recovery system (either hardware or software) in place to reduce transmission errors?

566. Is a regular check of transmission lines performed to detect for electronic surveillance devices, hacker intrusion equipment, or wiretaps?

VIRUS DETECTION AND PREVENTION

567. Are virus detection programs currently installed on the facility's computer systems?

568. Name the type and function of the computer virus detection programs which are currently installed on the computer systems.

569. Are the following areas checked for signs of virus activities by the selected virus detection program?

    A. .EXE files

    B. .OVL files

    C. .PIF files

    D. .SYS files

    E. .COM files

    F. Deliberately hidden files

    G. Boot sector

    H. RAM memory

    I. All memory addresses and occupying programs TSRs (Terminate-and-Stay-Resident)

    J. Partition tables

570. Does the virus protection software perform the following functions?

A. Examine all local and network drives for virus activity automatically

B. Develop a comparative program signature to detect any differences created by viruses

C. Use a checksum methodology of all executable programs for comparative analysis

D. Continually check critical sectors and executable programs for virus-related activity

571. Are all personal computer systems installed with a virus detection program, which is updateable?

572. If a virus is detected, what specific procedures are implemented?

573. Are these kept current in relation to current virus threats?

574. Are computer area personnel responsible for updating personal computer virus detection software?

575. Can the virus detecting software scan “ARC” or “ZIP” files without unarchiving them if it is a personal computer system?

576. Is exposure to boot viruses limited by not allowing the computer to be booted from the floppy drive without a password authorization access?

577. Is it prohibited to use substandard information sources such as diskettes brought from home or using a home computer dial-in modems on company computer systems, because these items may carry viruses?

578. Are employees not permitted to use company computer systems to communicate with any unauthorized and unapproved outside bulletin board systems because this may be a source of viruses?

579. Are the various types of viruses, such as worms, Trojan horses, logic bombs, etc., their characteristics, and their methods of infection and deletion all addressed in the computer security training program?

580. Are all the LANs protected by reliable virus detection software?

581. Is the version of the network virus detection software continually upgraded?

582. Do the virus detection programs help prevent, detect, and cure viruses in both the LAN and PC versions?

583. Does the virus detection program apply a write-protect designation on files and signal an alert if there are any alterations attempted to protected files by a virus?

584. Is any detected viral strain automatically cleaned or eliminated by the virus software?

585. What action does the virus detection program automatically initiate when a virus is detected?

586. Is the virus detection software's integrity checked by its own program features?

587. To protect its own files from viral damage, does the virus detection software have an automatic protection feature?

588. Are any attempts to write-over or reformat prevented by the virus detection software's own features?

589. Does the virus detection software simply deactivate viruses or does it remove any identified viruses discovered in memory?

590. Is installation of this program relatively easy?

591. Is important information routinely backed-up?

592. Are multiple backup copies made?

593. Are there controls that limit the access of most users?

594. Are only key computer area personnel allowed to have access to software and operating system programming?

595. Is the system and all its peripheral devices continuously monitored for any irregularities in operation?

596. Are users required to immediately report and document any irregularities?

597. Is all of the original software scanned for viruses and then write-protected?

598. After a viral attack, is only the virus-free, write-protected software reinstalled?

599. Is public domain software prohibited unless it has been thoroughly checked and tested on a separate standalone system?

600. When testing software for viruses, is there a procedure for using different future computer clock settings such as holidays or April 15th, etc.?

601. Is all of the software from bulletin boards prohibited unless it has been thoroughly checked and tested on a separate standalone system?

602. Is the source of all software known and verified before it is used?

603. Is the operating system software stored on virus-free, write-protected diskettes?

604. Are all commercially manufactured software checked to see if they are still sealed in their original packaging when purchased?

605. Is there restricted access and use of the original software diskettes?

606. Is there a policy prohibiting any copying of original software programs except those designated and approved backups needed by the computer area?

607. Is a periodic audit conducted on all software programs inventory and the corresponding user-request records?

608. Are the legal ramifications and penalties for illegal copying of software programs fully explained to all employees?

609. Does the company have a specific policy for taking action against employees who illegally copy software?

610. Does this policy contain specific disciplinary actions levied against violators?

611. Is the copying of software programs to portable storage media or PCs prevented by system controls and configurations?

612. Are all employees, especially programmers, not permitted to perform out-of-the-company programming services?

613. Are all disks formatted with a virus-free, write-protected original?

614. Are all files and programs periodically inspected to determine if there is an unexplained increase in their size?

615. Are all files and programs periodically inspected to determine if there is any unexplained changes in the dates of modifications?

616. Are any critical information files protected by individual file passwords?

617. Are all source codes routinely inspected for viruses, is it prohibited to share object codes?

618. Are reused diskettes overwritten with a "format" command rather than being "erased?"

619. On LANs, are diskless terminals used whenever possible?

620. Are all company software source programs labeled with a protective statement and a unique company identifier?

621. Before any program distribution occurs are all copies checked for the correct version and proper operation?

622. Does the backup system and current operating system interface even if they are not exactly the same?

623. Is the backup system and its software periodically tested?

624. When testing program performance, is specific information with known results used as a comparison?

625. Is the use of software that is dependent on a specific piece of equipment avoided?

626. Do all software license agreements allow for copies to be utilized on backup systems?

627. Is there a computer area person responsible for the tracking, control, storage, and security for all software programs?

TCP/IP

628. If **TCP** is being used, has **TCP-wrapper** been implemented, which allows an access-control list to restrict each network service supported (such as Telnet or **FTP**)?

629. Is **SYSLOG** enabled with "**auth.crit.**" parameter enabled?

630. Are network monitoring programs being actively utilized such as **NETLOG**, **COURTNEY**, **GABRIEL**, **ARGUS**, etc.?

631. When using Cisco routers, have **Loose source routing**, Finger, and **IP Forwarding** services been disabled, because of the fact that they can create vulnerabilities?

632. When using Cisco routers, have the most recent security patches been installed?

633. When using Cisco routers, is access controlled by "**Access-Class**"?

634. When using Cisco routers, has "**SNMP Write**" been disallowed?

635. When using Cisco routers, are usernames and passwords authenticated using **TACACS**?

636. When using Cisco routers, is the Access Control List routinely checked to ensure that any changes have not created any new security vulnerabilities? (A program like "**pollem**" can assist in this effort.)

UNIX (*May not apply to all versions of UNIX*)

UNIX Passwords

637. Have **shadow password** and **group** files been implemented to minimize access to encrypted passwords?

638. Have **passwd+** or **npasswd** been implemented to enforce more secure password rules?

639. Are passwords routinely checked by a password cracking program, such as **Crack**, to determine if any are weak?

UNIX Log-in

640. Has a log-in banner been implemented on your system so it appears to users before they log-in or immediately after they have logged-on?

641. Does the log-in banner identify who owns the system?

642. Does the log-in banner identify who is authorized to use the system?

643. Does the log-in banner state that unauthorized use of the system is illegal and violates the system owner's policy ?

644. Does the login banner inform users that their activity is being monitored, and what will be done with this gathered information?

Root Access

645. Are direct **root** logins not permitted, except may be from a designated console in a physically secure area?

646. Are designated consoles/terminals marked as secure in the **/etc/ttytab** file that will allow any user with **UID=0** to login into **root** directly?

647. Is the **/etc/ttytab** file (BSD UNIX) owned by **root**?

648. Does the **/etc/ttytab** file (BSD UNIX) have its permissions set at **644**?

649. Are all other terminals marked as unsecure?

650. In the **/etc/hosts.lpd** file, is the first character prohibited from being "**-**" ?

651. In the **/etc/hosts.lpd** file, are the permissions set to **600**?

652. In the **/etc/hosts.lpd** file, is the owner set to **root**?

653. Is the number of users who can **su** to **root** kept to an absolute minimum?

654. Is all **su** activity monitored and logged?

655. Can a system administrator give certain users limited **root** access and log the activity since a program like **sudo** has been implemented?

656. In the **/.profile** file, has "**.**" been removed from the path definition?

657. In C-shell **/.login** and **/.cshrc** files, has "**.**" been removed from the shell variable **path**?

658. Are all the directories specifically identified in **path** unable to be publicly or group writable?

659. If **NFS** is in use, does **root** own all executable programs and their associated directories to guard against a security weakness in **NFS uid-mapping**?

File System Security

660. Do all **.exrc** files have legitimate purposes?

661. Are **.forward** files in user home directories prohibited to execute a command or run a program?

662. Is the **/etc/rc.local file** prevented from **chmod 666 motd**?

663. When creating the **/etc/motd** file, is the command "**rm -f /tmp/t1**" (or similar) implemented in **/etc/rc.local** to clean up the temporary file? (This should run on the local daemons before the code at startup.)

664. Has the latest version or patches of **/usr/lib/expreserve** been implemented?

665. Are the **/etc/utmp** permissions set to **644**?

666. Are the **/etc/sm** and **/etc/sm.bak** permissions set to **2755**?

667. Are the **/etc/state** permissions set to **644**?

668. Are the **/etc/motd** and /**etc/mtab** permissions set to **644**?

669. Are the **/etc/syslog.pid** permissions set to **644**?

670. In the **/usr/kvm/crash** file, have the **setgid** privileges been removed?

671. If the kernel owned by **root** (e.g., **/vmUNIX**), has group been set to **0** (wheel on SunOS) and the permissions set to **644**?

672. Are **/etc**, **/bin**, **/usr/etc**, **/usr/bin** and **/tmp** files owned by **root** and has the "sticky-bit" been set on **/tmp**?

673. Are there no unexpected world writable files or directories on the system?

674. Are all files run by **root**, owned by **root**?

675. Are all files run by **root** not group writable or world writable?

676. Are all files run by **root** in a directory where **root** owns every directory in the path and is not world writable or group writable?

677. Are all **bin** owned non-**setuid** files, non-**setgid** files and directories, which are world readable but not world or group writable changed to **root** ownership?

Permissions

678. Is the "sticky bit" set on all publicly writable directories, which prevents users from deleting files they don't own (usually **/tmp**, **/usr/tmp**, and the publicly writable directories under **/usr/spool**)?

679. Have all **set-uid** programs been checked for excessive privileges?

680. Has a safe **uid** been given to "**nobody**" and "**nogroup**"?

681. In the **/etc/inetd.conf** file, has a "#" been placed in column 1 of each of the unused services to disable them?

682. In the **/etc/inetd.conf** file, are the permissions set to **644**?

683. In the **/etc/inetd.conf** file, is it owned by **root**?

684. In the **/etc/services** file, are the permissions set to **644**?

685. In the **/etc/services** file, is it owned by **root**?

686. In the **/etc/aliases** file, has a "#" been placed at the beginning of the line, which comments out the "decode" aliases?

687. In the **/etc/aliases** file, are all executable programs owned by **root**?

688. In the **/etc/aliases** file, are the permissions set at **755** and stored in a systems directory?

689. In the **/etc/aliases** file, is it owned by **root**?

690. Is **fsirand** running for all file systems?

691. Is **Ssh (Secure Shell**) being used when it is necessary for a computer to log-in to another computer, execute remote commands, and to move files?

692. On each UNIX system, have all world-writable **/etc/cshrc** or **/etc/profile** files been removed (use COPS program)?

693. On each UNIX system, have all default vendor accounts been removed (use COPS program)?

694. On each UNIX system, have all world writable files in system directories been removed (use COPS program)?

695. On each UNIX system, have all "**.**" in **root's path** been removed (use COPS program)?

696. On each UNIX system, has a **root** in **/.rhost** file been removed (use COPS program)?

697. On each UNIX system, has the correct format been used for **/etc/passwd** file (use COPS program)?

698. On each UNIX system, have all duplicate group names been removed (use COPS program)?

699. On each UNIX system, have all world-writable files in home directories been removed (use COPS program)?

700. On each UNIX system, have all world-writable user environment (dot) files been removed (use COPS program)?

701. On each UNIX system, have all programs called by **root**'s **.cshrc** and **.login** files been write-protected from nonowners or other users (use COPS program)?

702. On each UNIX system, are all programs executed by **cron** located in protected directories with the appropriate file permissions (use COPS program)?

703. On each UNIX system, have all disk drive device files been examined for the correct ownership and permissions (use COPS program)?

704. On each UNIX system, are all programs executed by the **/etc/rc*** file owned by **root** and not world writable (use COPS program)?

705. On each UNIX system, are all file systems located in the **/etc/exports** file prohibited from being world exportable (use COPS program)?

706. On each UNIX system, are all group-writable files authorized (use COPS program)?

707. On each UNIX system, are there any **root** environment files such as **aliases**, **umask**, and **.in path** (use COPS program)?

708. In addition to using the COPS program, is the **Tiger** program also used to identify ways that **root** can be compromised?

709. Is the **TFTP** daemon disabled unless it is absolutely necessary?

710. If **TFTP** daemon is enabled, has the "**-s flag**" been set?

711. If **finger** is enabled, is it protected by **TCP-wrapper** or a custom daemon?

712. If **systat** is enabled, is it protected by **TCP-wrapper**?

713. If **netstat** is enabled, is it protected by **TCP-wrapper**?

714. If **walld** is enabled, is it protected by **TCP-wrapper**?

715. As part of **NFS**, is **mountd** started from the **/etc/rc** directory?

716. Is the **rexd** mechanism disabled unless it is absolutely necessary?

717. If **httpd** is enabled, is it protected by **TCP-wrapper**?

718. Is the latest version of **Sendmail** running to help avoid known security holes?

719. Are there accounts which have no passwords but execute a command disabled (such as "**sync**")?

720. Are user account **.rhosts** files not permitted to contain local machine names?

721. Is a "**raudit**" program used to check for vulnerable **.rhosts** entries?

722. Have all **/.rhosts** files been removed unless absolutely necessary?

723. Does **ruserok( )** ignore **.rhosts** files that are group or world readable?

724. If not, have all **.rhost** files been removed from **root** unless absolutely necessary? Has **/.rhosts** been set so that no other machine is equivalent?

725. Have all **/etc/host.equiv** files been removed unless they are absolutely necessary?

726. Are all systems routinely scanned for known security vulnerabilities with a specialized software product designed for this purpose?

File Transfer Protocol (FTP)

727. Is the latest version of **FTP-daemon** running?

728. Does the **FTP**-server use an **FTP-daemon** with enhanced access control and logging capabilities?

729. Has the **FTP-server** been tested to ensure it does not have the **SITE EXEC** command?

730. Are any directories such as **~ftp/bin**, **~ftp/usr/bin**, or **~ftp/sbin**, which can execute **SITE EXEC**, prohibited from containing any system commands or include a shell?

731. Have all **rename**, **delete**, **overwrite**, **umask**, and **chmod** options (check for others) been denied for any anonymous users and guests if a configurable version of **FTP** is being used?

732. Does the **/etc/ftpusers** file on the **FTP**-server contain those users that are prohibited from connecting to **ftpd**? (At the very least, any vendor accounts, e.g., **root**, **bin**, **daemon**, **uucp**, **nobody**, **ingres**, **news**.)

733. In the system **passwd** file, is an invalid password and user shell used for the **FTP** entry?

734. Has the system been checked to ensure a copy of the real **/etc/passwd** file is not found as **~ftp/etc/passwd**?

735. Does the **~ftp/etc/passwd** file have permissions of **444** and not contain the names of any accounts in the real password file?

736. Does the **~ftp/etc/passwd** file only contain **root** and **ftp**? (These should be bogus entries with nonfunctioning passwords.)

737. Has the system been checked to ensure a copy of the real **/etc/group** file is not found as **~ftp/etc/group**?

738. Has the **~ftp/etc/group** file been created with the permissions of **444** and owned by **root**?

739. Do the **~ftp/.rhosts** and **~ftp/.forward** files have permissions of **600**? Are they owned by **root**? Have they been set to zero length?

740. Is the **FTP** account not permitted to contain any files or directories which are owned by the **FTP** account or have the same group as the **FTP** account?

741. Are any anonymous **FTP** users prevented from creating files or directories in any directory?

742. Are **FTP** users only allowed to read information in public areas?

743. Are the **FTP home directory** (~**ftp/**) permissions set to **555** (**read nowrite execute**) and owner set to **root** (NOT **ftp**)?

744. Do the system subdirectories **~ftp/etc** and **~ftp/bin** have the permissions of **111** only and owner set to **root**?

745. Do the permissions of files in **~ftp/bin/*** have only the permissions of **111** and the owner set to **root**?

746. Do the permissions of files in **~ftp/etc/*** only have the permissions of **444** and the owner set to **root**?

747. Does **/usr/spool/mail/ftp** only have the permissions of **400** and the owner set to **root**?

748. Have all **FTP** writable directories been eliminated?

749. If there is a need for an **FTP** writable directory, has it been limited to one?

750. If there is an **FTP** writable directory, has it been made to not be readable?

751. If there is an **FTP** writable directory, has its permissions been set to **1733** and is it owned by **root**?

752. If there are **FTP** writable directories, is it possible to have them on a separate partition?

753. Is the **comsat** daemon disabled unless it is absolutely necessary?

754. Is the **talk** function disabled unless it is absolutely necessary?

UNIX to UNIX Copy (UUCP)

755. Is **uucp** disabled unless it is absolutely necessary?

756. Have **uucp-related shell commands** which execute permissions been turned been disabled?

757. If **uucp** is in use, have all **.rhosts** files been removed from the **uucp** home directory?

758. Is the **L.cmds** file owned by **root**?

759. Have all world writable **uucp** files been removed?

760. Does each site which needs access to a local machine have a different **uucp** login?

761. Does each **uucp** login have only the bare minimum number of commands it can execute?

Network File System (NFS)

762. If **NFS** is not being used, has the **NFS** daemon been disabled?

763. If **NFS** is used to mount remote file systems, is it required to be mounted with the **nosuid** option (**-o nosuid**)?

764. If **NFS** is used to mount remote file systems, can it be mounted as **read-only** (**-o ro**)?

765. If **NFS** is used to export file systems, is it only done when absolutely necessary and only to hosts which require them?

766. If **NFS** is used to export file systems, is it only permitted to **fully qualified** host names?

767. If **NFS** is used to export file systems, are export lists limited to 256 characters?

768. If **NFS** is used to export file systems, are the specific hosts listed in the **/etc/exports** file?

769. If **NFS** is used to export file systems, is it possible to make the directories **read-only**?

770. If **NFS** is used to export file systems, is **port monitoring** turned on?

771. Are all system programs and system directories, from parent directories to **root**, owned by **root**?

772. Are **UDP** or **TCP** packets prevented from passing to **ports 2049** (**NFS** daemon) or **111** (portmapper) on the firewall?

773. Is the **NFS** server prevented from referencing itself by name or by the loopback address in its own exports file?

774. Is the **NFS** server running a portmapper that doesn't permit proxy access?

775. Have the most recent security patches been installed for all applicable programs?

776. Have all "**r**" commands been disabled unless absolutely necessary?

777. If any of the "**r**" commands are used, are ports **512**, **513**, and **514** filtered (blocked) by the router(s) so that outside users cannot exploit them?

AIX (*May not apply to all versions of AIX*)

778. Has **sysinfo** been enabled to determine system inventory?

779. Has **tcpchk** been enabled to determine who is using **.rhosts** and **.netrc** files and what is the status of the **tcp** subsystem?

780. Is the **nsfck** script being used to determine the **NIS** and **NFS** subsystems status?

781. Is the **nethwchk** used to list the connected interfaces?

782. Have direct logins to **root** been disabled?

783. Have the appropriate user password characteristics been defined in **/etc/security/user**?

784. Have passwords been scanned for inconsistencies by using run **/usr/bin/pwdck – y** ?

785. Have groups been scanned for inconsistencies by using run **/usr/sbin/grpck – y** ?

786. Are **/etc/group**, **/etc/security/group**, **/etc/passwd**, and **/etc/security/passwd** routinely examined?

787. Is **npasswd** used for proactive password checking?

788. Is **COPS** run on a regular basis?

789. Is **/etc/security/limits** used to set the systemwide default process limits?

790. Is the **NIS** domain name kept extremely confidential since the **passwd** file will not be protected by **shadow passwd**?

791. Are all **suid/sgid** programs set to **mode 511** minimum?

792. Have **system dirs** been set to **perm 755** minimum?

793. Are the **system dirs** used by **root** for programs owned by **root**?

794. Is the "sticky bit" set on **/tmp** and **/usr/tmp**?

795. Are general users prevented from having access to **UUCP**, **cron table**, and **syslogs** system source codes?

796. Is AIX default set at **555** for **/usr/sbin/mount**, **/usr/sbin/acct/acctcom**, **/usr/sbin/login**, and **login**?

797. Is the **/usr/bin/sysck** command used to monitor the checksums for **passwd**, **su**, **login**, **rlogin**, **tip**, **cu**, and **crypt**?

798. Does the **system dir** contain only one copy of system commands?

799. Does **root** own disk, storage, tape, and network device files and are the perms set at **600**?

800. Are the User HOME **dirs** set to mode **710**, not **711**?

801. Are the **.profile** and **.login** files set to mode **600** or **640**?

802. Has the appropriate **umask** been set in **/etc/profile**?

803. Has the use of the "**r-commands**" been restricted or even possibly disabled?

804. Have all '**at**' jobs owned by **root** had their function verified?

805. Are users prevented from using **rsh**, **rcp**, **rlogin**, and **TFTP** unless these have been secured?

806. Have **.rhosts**, **/etc/hosts.equiv**, or **.netrc files** been disabled?

807. Is **securetcpip** running?

808. Are **/etc/hosts** set to **660**?

809. Are **/etc/hosts.equiv**, **.netrc**, **/etc/hosts**, **.rhosts**, and **/etc/inetd.conf** regularly checked against a checklist version?

810. Is **netstat -p tcp** used to look for failed connection attempts?

811. Has the **FTP** entry in **/etc/inetd.conf** been changed to **ftpd -l** to enable logging?

812. Has AIX auditing been enabled?

813. Is the "**/usr/sbin/watch**" command used to observe audit events for an "untrusted" program?

814. Is a separate account set up to do software installs?

815. Is the **tsh** command used to run a trusted command interpreter?

816. Has file security been enhanced by using **/usr/bin/aclget**, **/usr/bin/acledit**, and **/usr/bin/aclput** as Access Control Lists?

817. Are failed logins tracked by **/usr/bin/who -s /etc/security/failedlogin**?

818. Is the **IFS** (input field separator) prohibited from being changed in **/etc/profile** and is its usage monitored in user profiles?

819. Are the default AIX file permissions set at **644** for files and **755** for directories?

820. If so, are user profile files set to **740** and system **dirs** have set other perms = **0**?

821. Is UUCP not used unless it is absolutely necessary?

822. Is the **/usr/bin/strings** command used to check any suspicious binaries?

823. Are copies of **syslogs** routed to a central machine?

Solaris (*May not apply to all versions of Solaris*)

824. Has the **/var/adm/loginlog** file been created and enabled?

825. Are kernel audit events frequently checked to ensure they have not been modified in an unauthorized manner?

826. Has the audit subsystem been enabled?

827. Has auditing been enabled for all users?

828. At a minimum, does auditing include the date and time of the event, **UserID**, type of the event, and success or failure of the event?

829. Does auditing for I&A events also include the origin of the request (e.g., terminal ID)?

830. Does auditing track events which might introduce an object into a user's address space?

831. Does auditing track events for the following?

    A. Object deletion events

    B. The name of the object

    C. MLS systems

    D. The object's security level

832. Are the audit data protected by the system, so that access to it is limited to those who are authorized to view audit data?

833. Does the system provide the capability to correlate all system, administrative, and audit logs?

834. Does the system provide the capability to use audit reduction?

835. Have any unauthorized changes been made to the **audit_warn** script?

836. Does the system provide the capability to perform system and database backups on a regular basis?

837. Has **cron** been securely configured?

838. Are all directories prevented from being world or group writable?

839. Is Administration Tool use limited to only authorized users?

840. Are each account's permissions set so that no user is able to write to any directory that should not be written to?

841. Is a record kept of every time the **su** command is used, who uses it, and when it is made in the log file, **/var/adm/sulog**, so that the system administrator can track who is using the superuser account?

842. Is the lockout function available for users to manually invoke?

843. Are all users required to reauthenticate themselves to unlock a locked terminal?

844. Does the system protect objects from unauthorized access?

845. Is the system capable of including or excluding access to each object on a per user and on a per group basis?

846. Is the system capable of restricting access to objects based on the user's identity and on access modes such as read, write, and execute?

847. Does the system's shell reset the Internal Field Separator (IFS) variable when invoked?

848. Is the **expreserve** executable secure?

849. Has **IP** forwarding and **IP** source routing been disabled?

850. Are none of the directories in the **root** search path world writable?

851. Are **root**'s startup files only writable by **root**?

852. Are all **root** executable files owned by **root** and not world or group writable?

853. Do all of the world-writable files on the system have a legitimate need for world-write access?

854. Are all world-readable, but not world or group writable, non-**setuid/setgid** system files and directories owned by **root**?

855. Are the startup and shutdown scripts valid and protected?

856. Do all of the **SUID** and **SGID** files on the system need their **SUID** and **SGID** privilege?

857. Is a general user prevented from changing the ownership of an **SUID** or **SGID** file (or any file) to any other user especially **root**?

858. Have all unauthorized device files been removed from the system?

859. Is **Finger** disabled unless there is a legitimate need for it?

860. Does the **ftpusers** file contain a list of users who are not allowed access to the system using the **FTP**? (If this file is missing, the list of users is considered to be empty, so that any user may use **FTP** to access the system if the other criteria for access are met.)

861. Is **TFTP** disabled unless there is a legitimate need for it? (The use of **TFTP** does not require an account or password on the remote system.)

862. If it is running, is the **TFTP** file prohibited from having the **SUID** or **SGID** bits set?

863. Is **Anonymous FTP** disabled unless there is a legitimate need for it?

864. Is the user challenged for the **eeprom** password when booting into single-user mode?

865. Are **I&A** mechanisms configured for secure operation?

866. Are permissions for the **/etc/security/passwd.adjunct** file set at **600**? Is the owner **root** showing that access to this file is limited to the owner (**root**)?

867. Is any user who tries to remotely log into the system first required to login to his/her account and then use the **su** command to become **root**?

868. Have all accounts on the system which have not been used within a reasonable amount of time been removed?

869. Have all duplicate **GID**s been eliminated?

870. Are all passwords which are transferred across the network protected?

871. Are all users on the system required to have a password?

872. Are the installation-provided **UserID**s checked to ensure they do not have default passwords?

873. Is password life limited to a maximum of 30 days? Is the user is notified prior to password expiration?

874. Is the Operating System designed to satisfy the **C2** level of trust as defined in the "Orange Book"?

875. Are there any duplicate **UID**s?

876. Are passwords stored in special files (shadow password files) that cannot be read by most users on the system?

877. Have the **decode** and **uudecode** aliases been removed from the aliases file (**/etc/aliases or /usr/lib/aliases**)?

878. Does **Sendmail** not respond to the **wiz** command? (If it does, **Sendmail** needs to be upgraded immediately.)

879. Does **Sendmail** not respond to the **debug** command? (If it does, **Sendmail** needs to be upgraded immediately.)

880. Does **Sendmail** not respond to the **kill** command? (If it does, **Sendmail** needs to be upgraded immediately.)

881. Has the latest version of **Sendmail** and any security patches been installed?

882. Has the **Sendmail -d bug** been eliminated?

883. Is a security warning displayed prior to the login process indicating restrictions that apply to logins, the highest classification of information processed on the system, and that misuse is subject to applicable penalties?

884. Are the network services configured securely (checked by browsing the **/etc/inetd.conf** file)?

885. Are the permissions of the **/etc/inetd.conf** file set at **600** and owned by **root**?

886. Are the permissions of the **/etc/inet/services** file set at **600** and owned by **root**?

887. Has the **.netrc** file been eliminated?

888. Are subnet addresses appropriately configured?

889. Have the correct subnet definitions only been obtained from the local network administrator?

890. Is **Cron** used to periodically check for, report the contents of, and remove **.rhosts** files?

891. Because the **NFS** server maps **root** to nobody, are files and directories protected on the server by setting their owner to **root** and making them not world writable or group writable?

892. Are file systems (**NFS**) mounted with the **nosuid** option and read-only where practical? (If read-only is not practical, verify system files and user home directories which are not mounted.)

893. Does the file **/etc/dfs/dfstab** have permissions of **644** and is owned by **root**?

894. Has **NFS** port monitoring been enabled?

895. Is **fsirand** run for all file systems and periodically re-run (part of **NFS** port monitoring)?

896. Is the **RPC portmapper** not permitted to allow proxy requests (part of **NFS** port monitoring)?

897. Is **rusers** disabled unless there is a legitimate business need for it?

898. Are only local hosts allowed in the **hosts.equiv** file?

899. Are all nonlocal hosts (including hosts at remote sites of the same organization) and any machines that are installed in "public" areas not trusted (in the **hosts.equiv** file)?

900. Are unauthorized interfaces not permitted to run in promiscuous mode (Ethernet sniffers)?

901. Are object reuse provisions enforced by the operating system and/or by features in the application software?

902. Is the file **/etc/logindevperm** owned by **root**? Does it have permissions of **644**?

903. Are the development tools such as language compilers, linkers, and debuggers adequately protected and only accessed by authorized users?

904. Are the default account initialization files secure (**/etc/profile** file)?

905. Are the default account initialization files owned by **root**? Do they have permissions of **644**?

906. Are the window tool scripts appropriately configured?

907. What operating system modules have been installed on the system?

908. Have the appropriate operating system patches been applied?

909. Is the system kernel configuration file correct (been secured)?

910. Are only authorized local and remote printers defined in the **/etc/lp/printers** directory?

911. Are unprivileged users prevented from executing the **crash** program?

912. If the file size of **/tmp/file1** is 0, will the **telnet daemon** be immediately replaced (telnet bug)?

913. Are users always prompted for a password when **telneting** into a host machine?

914. Does the **uucp** system not allow a command to be encapsulated in addresses to prevent system execution of commands encapsulated in addresses?

915. If **uucp** is being used, are the **uucp** programs owned by **uucp** and not by **root**? Have the proper permissions been set?

916. Are the **uucp** programs running **SUID uucp**, not **SUID root**?

917. If the **uucp** entry is enabled, does the **/etc/uucp/Permissions** file only allow minimal access? An empty Permissions file provides minimal access.

918. Is the **uucp** home directory not in a directory that is world writable?

919. Is the home directory for the **uucp** account not in the directory **/usr/spool/uucp/uucppublic** or any other directory that can be written to by a **uucp** user?

920. Is the **http** server daemon not being run as **root**, but as a specially created nonprivileged user such as **httpd**?

921. Are the file permissions on all **http** clients listed not owned by **root** and are not **SUID**?

922. Does the system use the **xauth X server** access control mechanism instead of the **xhosts** mechanism?

923. Are the **setuid** and **setgid** privilege bits not set on the **xterm** program?

924. If possible, is **xhost** security not used and is **xauth** or another security mechanism used?

925. Is the **X server** vulnerable to denial of service attacks?

HP-UX

926. Is the audit subsystem configured correctly and securely?

927. Is auditing enabled?

928. Is auditing required to collect audit events such as login and logout, use of privileged commands, application, and session initiation?

929. Is auditing required to collect audit events for the use of print command and **DAC** permission modification?

930. Is auditing required to collect audit events on the export to media?

931. Is auditing required to collect audit events for unauthorized access attempts to files?

932. Does the system provide the **SGSO** with the capability to select and enable audit events including use of **I&A**, introduction of objects into a user's address space, deletion of objects, trusted user actions, print use, etc.?

933. Is the "audit all users" parameter selected?

934. Are the required parameters identified for each recorded audit event including date and time of event, **UserID**, type of event, success or failure of event, I&A events, the origin of the request, etc.?

935. Is the audit data protected by the system so that access to it is limited to only those authorized to view the audit data?

936. Does the system provide an auditing function capable of accepting application level audit logging requests? Is a standard audit format provided for use in application level auditing?

937. Is the audit data protected from change or deletion by general users?

938. Does the system provide the capability to perform system and database backups on a periodic basis?

939. Are the **cron.allow** and **cron.deny** files owned by **root**, not world writable, and do they contain the correct entries?

940. Is the lockout function available for users to manually lock their terminals? Are users required to reauthenticate themselves to unlock a locked terminal?

941. Is the system capable of restricting access to objects based on the user's identity and on access modes (e.g., read, write, execute)?

942. In the file system security, does the system reset the IFS variable and are **SUID** or **SGID** scripts never used?

943. Is a **setuid root Bourne shell** prohibited from being in your home directory?

944. Is **root**'s search path prevented from any directories that are writable by other users?

945. Is the file system appropriately partitioned so that no file system is approaching 100% full?

946. Are **root**'s startup files permissions **400** or **600** and only writable/owned by **root**?

947. Are all executables run by **root** located in a directory where every directory in the path is owned by **root** and is not group or world writable?

948. Is any directory that contains a script, executable, or configuration file that is executed in the **rc** scripts during bootup and shutdown not writable by a user other than **root**?

949. Is any device outside the **/dev** and **/devices** directories, not owned by **root**, prohibited?

950. If **finger** is enabled, has it been securely configured?

951. Does the **ftpusers** file contain a list of users who are denied access to the system using **ftp**?

952. Has the **-s** option for **tftpd** been enabled and set to /**tftpboot** so that it will only start with the **home** and the **root** directory?

953. Does the **TFTP** file not have the **SUID** or **SGID** bits set?

954. If anonymous **FTP** is enabled, is the directory **ftp/bin** owned by the superuser and unwritable by anyone?

955. Have all accounts on the system that have not been used within a reasonable amount of time (30 days) been disabled?

956. Has the system been checked to ensure there are no duplicate **GID**s?

957. Does the system enforce individual user accountability and a globally unique valid **UserID**?

958. Does the system enforce that a password is required for all users to access the system?

959. Does the system enforce that the user's identity be associated with all audit actions performed?

960. Have the default passwords been changed on the installation-provided **UserIDs**?

961. Is password life limited to 30 days? Is the user notified prior to password expiration?

962. Do only authorized usernames appear with the **root** group name?

963. Has the system been checked to ensure that there are no duplicate **UIDs**?

964. Is site-identifying information stored for all user accounts on the system?

965. Have all guest accounts been disabled?

966. If guest accounts do exist, are they protected by a strong password (a minimum of eight characters using a combination of letters, numbers, and symbols)?

967. Is authentication data protected from being accessed by unauthorized users?

968. Have the "decode" and "uudecode" aliases been removed from the aliases file (**/etc/aliases** or **/usr/lib/aliases**)?

969. In **Sendmail**, has the **wiz** command been removed?

970. In **Sendmail**, has the **debug** command been removed?

971. In **Sendmail**, has the **kill** command been disabled?

972. Have the latest patches for **Sendmail** been installed, and are they properly configured?

973. Are all programs executable by an alias owned by **root** have permissions **755** and stored in a systems directory (e.g., **/usr/local/bin**)?

974. Has **Sendmail** been checked to ensure that it does not contain **-d bug**?

975. Is a security warning displayed prior to the login process indicating restrictions that apply to log-ins, the highest classification of information processed on the system, and that misuse is subject to applicable penalties?

976. Are all unnecessary network services disabled?

977. Has the **.netrc** file been removed as part of making a more secure system?

978. Have the correct subnet definitions been obtained from the local network administrator?

979. Is "**-**" not permitted to be the first character of any **.rhosts** file?

980. Are the permissions of all **.rhosts** files set to **600**?

981. Is the owner of each **.rhosts** file the account's owner?

982. Are any **.rhosts** files not permitted to have the symbol "+" on any line?

983. Does the usage of netgroups within **.rhosts** not allow unintended access to this account?

984. Are any of the **.rhosts** files not permitted to use "**!**" or "**#**"?

985. Are the **filesystems** mounted with the **nosuid** option and read-only where practical?

986. If read-only is not practical, are system files and user home directories not mounted?

987. Have all **NFS** patches been applied?

988. Is there a periodic review of what you currently have exported?

989. Is **fsir** activated for all your file systems?

990. Is the **RPC portmapper** prohibited from allowing proxy requests?

991. Has the "+" line in the **/etc/passwd** file for each **NIS** client been correctly entered?

992. Does a local host's **ypbind** not accept **ypbind -H** commands from remote hosts?

993. Is **rusers** disabled unless there is a legitimate business need?

994. Does **root** own **/etc/hosts.equiv**?

995. Are the permissions of /**etc/hosts.equiv** set to **600**?

996. Has the first character of /**etc/hosts.equiv** been checked to ensure that it is not "-"?

997. Does /**etc/hosts.equiv** not contain a line with only a "+" (a plus sign)?

998. Does **/etc/hosts.equiv** list only a small number of trusted hosts? Are all hosts listed within your domain or under your management?

999. Does **/etc/hosts.equiv** not include "**!**" or "**#**"?

1000. Are all hosts in **/etc/hosts.equiv** specified using **IP** addresses to mitigate **DNS** spoof attacks?

1001. Are **netgroups** in **/etc/hosts.equiv** used for easier management?

1002. Has the system been checked to ensure that no interfaces are in promiscuous mode (Ethernet)?

1003. In **/etc/ttys**, is only the operator's console defined as "secure"?

1004. Are development tools such as language compilers, linkers, and debuggers adequately protected and only accessed by authorized users?

1005. By default, does the **/etc/profile** file set the user terminal type, check for new e-mail, and set the proper **umask (027 or 077**)?

1006. Are all user account initialization files owned by the user (or **root**) and have permissions of **640**?

1007. Are the default account initialization files owned by **root** and have permissions of **644**?

1008. Is the system kernel configuration file correct?

1009. Have the appropriate operating system patches been applied?

1010. Are the only appropriate printers defined?

1011. Has the system been tested to ensure that the **telnet bug** does not exist?

1012. Has the correct operating system version been installed?

1013. Is a user always prompted for a password when **telneting** into the host machine?

1014. Are all entries in the **lpd** file appropriate (i.e., only hosts that should access the local printers are listed)?

1015. Is **uucp** disabled if possible?

1016. Does the **uucp** system not allow a command to be encapsulated in addresses to prevent system execution of commands encapsulated in addresses?

1017. Do the **uucp** programs run **SUID uucp** and not **SUID root**?

1018. Other than being able to read the spooled **UUCP** files, is the **uucp** user not permitted to have any special privileges?

1019. Is **USERFILE** owned by **uucp** and not world writable?

1020. Is **L.cmds** owned by **root** (not by **uucp**)? Is it world readable?

1021. Is **uucp** callback enabled for enhanced security (Version 2 **uucp**)?

1022. Is the home directory for the **uucp** account not in the directory **/usr/spool/uucp/uucppublic** or any other directory that can be written to by a **uucp** user?

1023. Is the **http** server daemon not being run as **root**, but as a specially created nonprivileged user such as "**httpd**"?

1024. Does the system use the **xauth** X server access control mechanism instead of the **xhosts** mechanism?

1025. Is the **xterm** program not **SUID** or **SGID**?

1026. Is **xterm** logging not permitted?

1027. Has the X server been tested for a denial of service vulnerability?

Administration

1028. Are there established policies and procedures for adding and deleting users?

1029. Are there established policies and procedures for adding and deleting network addresses for servers, workstations, printers, etc?

1030. Are there established policies and procedures for network address naming conventions?

1031. Are there established policies and procedures for remaining in compliance with all software licensing agreements?

1032. Is an automated network inventory at the node level taken for all hardware at least twice a year?

1033. Is an automated network inventory at the node level taken for all software at least twice a year?

1034. Is there a network map/diagram which is current?

1035. Is automated network mapping performed every 3 months?

1036. What network architectures are currently being used (Ethernet, Token Ring, FDDI, etc.)?

1037. What file access control methods are being utilized?

1038. What type of WAN connectivity is being utilized between remote sites (e.g., T1, T3, Frame Relay, ATM, etc.)?

1039. What type of connection is the company's access to the Internet?

1040. Is a screening router used as the company's Internet access gateway?

1041. Which carrier or ISP (Internet Service Provider) is providing Internet access?

1042. Does the Internet connectivity provider offer any type of enhanced security measures?

1043. Does the Internet connectivity provider have any knowledgeable security staff?

1044. What liability does the Internet connectivity provider assume for security breaches?

1045. Is the Internet connectivity provider prepared to handle a "denial of service" attack?

1046. Is the system regularly audited for dormant accounts?

1047. Are dormant accounts disabled if they have not been used for a specified period of time such as 30 days?

1048. Are account renewal notices sent out to dormant account users? Are those accounts deleted if there is no response back?

1049. Do all accounts require and have passwords?

1050. Are any user areas routinely backed-up and archived?

1051. Are logs regularly monitored for successful and unsuccessful "superuser/administrator" attempts?

1052. Are logs regularly monitored for repeated login failures?

1053. Are logs regularly monitored for repeated **login refused** messages?

1054. Are special accounts not permitted to share accounts other than **root**?

1055. Are special groups used to restrict which users have access to **root**?

1056. Are accounts which have no passwords but execute a command disabled (such as "**sync**")?

WINDOWS NT

1057. Is Windows NT the only operating system installed on the machine?

1058. Are the subsystems **POSIX** and **OS/2** not installed?

1059. Is **NTFS** (NT File System), not the FATS file system, used to format all drives?

1060. Is the security log not permitted to overwrite any previous events unless the log is cleared manually?

1061. Is the use of blank passwords not permitted?

NT Audit

1062. Has the audit subsystem been enabled?

1063. Is audit correctly configured and does it collect the required audit events (log-in and log-out, use of privileged commands, application and session initiation, use of print command, DAC permission modification, export to media, etc.)?

1064. Are significant changes to selected Registry keys audited?

1065. At a minimum, are the following audit events audited?

A. Log-in (unsuccessful and successful) and Log-out (successful)

B. Use of privileged commands (unsuccessful and successful)

C. Application and session initiation (unsuccessful and successful)

D. Use of print command (unsuccessful and successful)

E. Discretionary access control permission modification (unsuccessful and successful)

F. Export to media (successful)

G. Unauthorized access attempts to files (unsuccessful)

H. System startup and shutdown (unsuccessful and successful)

1066. Are all unsuccessful log-in attempts logged?

1067. Is the system capable of detecting when the audit file reaches a configurable threshold and are audit records not lost if this threshold is reached? (If the audit file becomes full, verify the system is shutdown until the audit data is archived.)

1068. Is the security (audit) log directed to a large, reliable storage device?

1069. Does an audit reduction capability exist? (This capability can be either OS provided or an add-on product.)

1070. Is the audit data protected by the system so that access to it is limited to those who are authorized to view audit data?

1071. Is the audit function protected from change or deletion by general users?

1072. Is the right to manage auditing and the security log protected?

1073. Has a current emergency repair disk been created, updated, and is it protected with an appropriate password?

1074. Are DAC mechanisms used to restrict access by general users to input/output (I/O) devices, such as floppy disks and tape drives, and the capability to specify which users may access I/O devices?

Access Control

1075. Is the floppy not shared over the network?

1076. Is the CD-ROM not shared over the network?

1077. Is a “deadman” function provided that locks the user's terminal if user input devices have been idle for longer than a configurable time period?

1078. Is the configurable time period default to 5 minutes?

1079. Can any user input device be used to restore a locked terminal?

1080. Is the specific input value (whether from keyboard, mouse, or other pointer) used to activate the function that restores the locked terminal ignored except to activate the function?

1081. Is the system locked through a password-protected screen saver when left unattended?

1082. Are the System Administration Tools configured securely? Is their use limited to appropriate users?

1083. Is the Operating System designed to satisfy the **C2** level of trust as defined by the **TCSEC**?

1084. Do the users and groups on the system have the appropriate (least) privileges?

1085. Is it prohibited for users, even "Administrators," to have the "Debug programs" right? (This right should not be enabled if the system is a production system.)

1086. Are third-party services run under an account in which rights have been tailored to include only those rights essential to allow the service to perform its function?

1087. Is no service permitted to run under the "Administrator" account?

1088. Does no user, not even Administrator, have the "Logon as a service" right?

1089. Are the "**TEMP**" or "**TMP**" directories used by many applications as a repository for temporary files containing data protected from access by unauthorized users?

1090. Have the **ACL**s on all Registry keys (used to support applications not needed by "guest") been set to deny write and execute access for the group "Everyone," but have granted the "write" and "execute" accesses for the group "Users"?

1091. Have **IP** forwarding and **IP** source routing been disabled?

1092. Is **NTFS** the only file system type list on all hard drives?

1093. If FAT file system is necessary, is the "Secure System Partition" selected? (This will restrict access of the FAT boot partition to "Administrators" only and protects the system files on the FAT boot partition.)

1094. Are the permissions on each directory set to the following?

A. **Administrators — Full Control**

B. **Creator Owner — Full Control**

C. **Everyone — Add and Read**

D. **System Operators — Add and Read**

E. **System — Full Control**

1095. In the "**C:\Boot.ini**", "**C:\Ntdetect.com**", and "**C:\Ntldr**" files, are the file permissions set to **Administrators - Full Control** and **SYSTEM - Full Control**?

1096. In the "**C:\AUTOEXEC.BAT**", and "**C:\CONFIG.SYS**" files, are the file permissions set to **Everyone – Read**, **Administrators - Full Control**, **SYSTEM - Full Control**?

1097. In the "**C:\WINNT**" directory, are the directory permissions set to **Administrators - Full Control**, **CREATOR OWNER - Full Control**, **Everyone – Read**, **SYSTEM - Full Control**, **Users - Change**?

1098. In the "**C:\WINNT**" directory, are all "*.ini" file permissions set to **Administrators - Full Control**, **CREATOR OWNER - Full Control**, **Everyone – Read**, **SYSTEM - Full Control**, **Users - Change**?

1099. In the "**C:\WINNT\SYSTEM**" directory, are the directory permissions set to **Administrators - Full Control**, **CREATOR OWNER - Full Control**, **Everyone – Read**, **SYSTEM - Full Control**?

1100. In the "**C:\WINNT\SYSTEM32**" directory, are the directory permissions set to **Administrators - Full Control**, **CREATOR OWNER - Full Control**, **Everyone – Read**, **Server Operators – Change**, **SYSTEM - Full Control**?

1101. In the "**C:\WINNT\SYSTEM32**" directory, are the directory permissions set to **Administrators - Full Control**, **CREATOR OWNER - Full Control**, **Everyone – Read**, **Server Operators – Change**, **SYSTEM - Full Control**?

1102. In the "**C:\WINNT\SYSTEM32\CONFIG**" directory, are the directory permissions set to **Administrators - Full Control**, **CREATOR OWNER - Full Control**, **Everyone – List**, **SYSTEM - Full Control**?

1103. In the "**C:\WINNT\SYSTEM32\SPOOL**" directory, are the directory permissions set to **Administrators - Full Control**, **CREATOR OWNER - Full Control**, **Everyone – Read**, **Print Operators – Full Control**, **Power Users – Change**, **Server Operators – Full Control**, **SYSTEM - Full Control**?

1104. In "**C:\WINNT\SYSTEM32**" directory, has it been checked to ensure that the "**OS2**" directory does not exist?

1105. In the shared directory used as a central repository for user profiles (normally "**\<SYSTEMROOT>\WINNT\Profiles**), are the permissions set to **Administrators - Full Control**, **CREATOR OWNER - Full Control**, **SYSTEM - Full Control**, **Everyone - Add**?

1106. If the machine is a server, domain controller, or if its availability is otherwise critical, is it prevented from being shutdown without first logging on? (No user or group can shut down the system remotely.)

1107. In the "**C:\WINNT\REPARI**" directory, are the directory permissions set to **Administrators - Full Control**?

1108. In the "**C:\WINNT\SYSTEM32\NTBACK-UP.EXE**" file, are the file permissions set to **Administrators - Full Control** and **SYSTEM - Full Control**?

1109. Are the permissions on all executable files set to **CREATOR OWNER - Full Control**, **Everyone – Read**, **System - Full Control**, **Administrators - Full Control**?

1110. Do directories containing executable files deny write permission to all users not specifically authorized to maintain the executables contained in the directory?

1111. In the "**C:\WINNT**" and "**C:\WINNT\SYSTEM32**" directory/subdirectory, are the permissions set to **Administrators - Full Control**, **CREATOR OWNER - Full Control**, **SYSTEM - Full Control**, **Everyone - Read**?

1112. Can only privileged users shutdown, reboot, or restart a system (either locally or remotely)?

1113. Is the single user boot or system firmware password set and the system configured so that a password must be entered to boot to a single-user state?

1114. Are critical systems physically protected so that attackers cannot replace **BIOS**, drain the **BIOS** battery, or carry away disk drives?

1115. Is all removable media stored in a physically secure location? (System Administrators and users must be aware that data on floppy disks and backup media is NOT protected by file system security.)

1116. Do all users with accounts that are members of the "Administrators" and/or Domain Administrators group(s) have a second account that is not a member of either group? Are they trained to use it when not engaged in system administrative work?

1117. Does the system provide the capability to restrict multiple login failures?

1118. If the number of login failures reaches a configurable threshold (0 through 5), is the **UserID** locked and is the user prohibited from further login attempts?

1119. If a **UserID** is locked, is a notification sent to the appropriate person?

Passwords

1120. In the Account Policy's "Lockout Duration" dialog box, is the "Forever" checked to ensure that the System Administrator must intervene? (This enforces administration awareness of password-guessing attacks.)

1121. Are passwords complex (a mix of upper and lowercase letters and numbers or symbols)? Can the Administrator account be locked out from network logon?

1122. Is individual accountability enforced by providing the capability to uniquely identify each individual system user?

1123. Are users required to identify themselves before beginning to perform any actions that the system is expected to mediate?

1124. Is each user identified by a globally unique user name or **UserID** that follows a standard set of processes or rules for formation?

1125. Does the system provide the capability of associating the user's identity with all auditable actions taken by that individual?

1126. Does the guest account have a strong password making it difficult for intruders to access the system? (This should be done if some applications depend on the availability of "guest" and cannot be rewritten not to depend on the guest account.)

1127. Is a protected mechanism (e.g., passwords) used to authenticate each user's identity?

1128. Are passwords at least eight alphanumeric characters in length?

1129. Does the system provide the capability for users or the system to generate passwords only in accordance with specified selection rules?

1130. Do passwords have a maximum password age and minimum password age?

1131. Does the password character set consist of letters, numbers, and special characters?

1132. Are passwords prohibited from containing repeating characters (e.g., ee) or from using a dictionary words?

1133. Is password life limited to 30 days?

1134. Are users notified 5 days prior to password expiration?

1135. Has the "Minimum Password Age" been set to at least one day? (This makes it more difficult for users to bypass the no-reuse restriction.)

1136. If a shared **UserID** must be permitted, are mandatory profiles configured for shared **UserID**s?

1137. Are users required to log-on to change their passwords?

1138. Has the name of the administrator account been changed from "Administrator"? Is it kept secret from nonprivileged users?

1139. Is site-identifying information stored for all user accounts on the system?

1140. Is any system on which the "guest" account is enabled (with or without a password), isolated as much as possible from the rest of the network and is not trusted by other systems on the network?

1141. Is any system on which the "guest" account is enabled fall outside the Windows NT domain as well as being required to have unique user accounts?

1142. Is a security warning displayed prior to the login process indicating restrictions that apply to logins, the highest classification of information processed on the system, and that misuse is subject to applicable penalties?

NT Permissions

1143. Are anonymous **FTP** connections not permitted?

1144. If anonymous **FTP** is permitted, is it properly restricted (with a password) and secured?

1145. If **FTP** is permitted, is it properly restricted (with a password) and secured?

1146. Are all "**r**" commands disabled except for those specifically required?

1147. Is **DHCP** not used for **C2** configurations?

1148. Is **WINS** not used for **C2** configurations?

1149. Are only authorized users able to perform administrative tasks that can affect system security?

1150. Have the appropriate operating system patches been applied?

1151. Are only appropriate printers defined?

1152. Does the Security Services (or like function groups) maintain a domain for their own execution that protects them from external interference or tampering (e.g., by modification of their code or data structures)?

1153. If the site has many Windows NT workstations, is the site configured as a domain, a primary domain controller, and a backup domain controller, if available?

1154. Are security support tools provided to periodically determine the security posture of systems, to validate the strength of the authentication mechanism, and to determine changes to designated systems and application files?

1155. Is a current virus protection program specific for Windows NT and capable of checking for macro viruses available on the system?

1156. Are the permissions on "**HKEY_CLASSES_ROOT**" and all its subkeys set to **Administrators - Full Control, CREATOR OWNER - Full Control, Everyone - Read**, and **System - Full Control**?

1157. Are the permissions on "**HKEY_USERS\.DEFAULT\UNICODE**" Program Groups [all subkeys] set to **Administrators - Full Control, Everyone - Read**, and **System - Full Control**?

1158. Are the permissions on "**HKEY_LOCAL_MACHINE\Software\Microsoft\RPC\[all subkeys]**" set to **Administrators - Full Control, System - Full Control, CREATOR OWNER - Full Control**, and **Everyone - ONLY Query Value, Enumerate Subkeys, Notify**, and **Read Control**?

1159. Are the permissions on "**HKEY_LOCAL_MACHINE\Software\Microsoft\ WindowsNT\ Current Version[all subkeys]**" set to **Everyone - ONLY Query Value, Enumerate Subkeys, Notify**, and **Read Control**?

1160. Are the permissions on "**HKEY_LOCAL_MACHINE\Software\Windows3.1 MigrationsStatus\Current Version[all subkeys]**" set to **Everyone - Read**?

1161. Are the permissions on "**HKEY_LOCAL_MACHINE\Software\Microsoft\ WindowsNT\ CurrentVersion\Profile**" List set to **Administrators - Full Control, SYSTEM - Full Control, CREATOR OWNER - Full Control**, and **Everyone - Special Access (Query Value, Create Subkey, Enumerate Subkeys, Notify**, and **Read Control**?

1162. Is Windows NT the only operating system installed on the system?

1163. Is the system prohibited from having multi-boot capability from different directories (other than "**C:**")?

NT Accounts

1164. Is a maximum password age such as 30 days enforced?

1165. Are passwords required to be unique from previous ones?

1166. Can passwords only be changed when users are already legitimately logged on to the system?

1167. Are users locked out after three failed login attempts?

1168. After users are locked out, are they required to contact the system administrator to restore their account?

1169. Are these failed attempts logged and monitored?

1170. Are remote users' server connections automatically terminated at a specified time to minimize after-hours threats?

1171. Are former employee accounts removed as soon as possible?

1172. Are all users in groups in which they are given the minimum amount of rights and privileges to perform their job functions?

1173. Are user groups routinely checked for any excessive or unnecessary rights and privileges?

1174. Do groups also have specific day and time restrictions for system access?

1175. Are groups restricted to designated computers?

1176. If used, do all administrative and temporary accounts automatically expire?

1177. If users have dial-in capability, have the call-back options been specified for each user's telephone number? (Call back should not be relied on as the only means of user authentication.)

1178. Is each global group carefully examined to determine if they only have access to specified resources and domains?

1179. Have all folders and files been examined to determine if only those authorized groups can access them? (Try using a program like DumpACL to assist with this effort.)

1180. Are management group members routinely examined to remove any inactive or unnecessary accounts? (These groups include print operators, backup operators, account operators, server operators, and administrators.)

1181. Do all administrative users have an account for administrative functions and one for all other tasks?

1182. Are administrative users to use their administrative account only when required?

1183. Is the administrator password and account name changed every time a new system administrator takes over?

1184. Is the administrator account renamed an obscure name to make it more difficult to identify?

1185. Is a "dummy" administrator account (disabled) set up with an obvious administrator-type name to draw attacks to it, rather than to the real administrator account?

1186. Are there any accounts with unnecessary administrator privileges?

1187. Have all unauthorized and unnecessary users been removed from the Domains Administrators and Administrators groups?

1188. If the network has multiple administrators, are their job functions routinely reviewed to justify their need for administrator-level privileges?

1189. Has a special emergency administrator account been created (obscure name), and does it require a three part password to which three separate trusted people are each given one part? (This is done for emergency purposes to access the administrator account and it requires all three designated individuals to be present in order to use it.)

1190. Is the "Access this computer from network" right been blocked for the Administrators group?

1191. Is the "Access this computer from network" right been blocked for the "Everybody" group and selectively added back for those accounts who only truly need it?

1192. Is the Domain Administrator group access removed from a Windows NT Workstation when adding it to the network, and is only a specific Administrator account allowed to be added?

1193. Has the "**IUSR_computername**" guest account been removed from all Microsoft Internet Information Servers (IIS Web server)?

1194. Have the file extensions **.CMD** and **.BAT** been disabled for external **CGI** scripts in the Microsoft IIS Web server file mapping feature?

1195. Has the Web server been checked to ensure that it does not allow read and write access to any file in the server's document **root**?

1196. If Windows NT 4.0 is being used, has the **Rollback.exe** file been removed from **Support\Deptools** since it can delete critical system information?

1197. Are Restore files, backup files, and directories access rights only given to those users who truly need them to perform their job functions?

1198. Are only administrators allowed to log-on directly to servers?

1199. Are all security and audit logs only accessible to designated administrators, not all of them?

1200. Are audit logs structured to provide specific types of information, such as exception transactions or events, rather than all available information?

1201. Are only administrators allowed to take ownership of objects and files?

1202. Have all advanced rights been examined to determine that a user doesn't have inappropriate rights?

1203. Have all folders and files been carefully examined to determine if all user and group access to these resources is justified?

1204. Have all shared folders and share permissions been carefully examined to determine if all user and group access to these resources is justified?

1205. Have separate folders been created for data files and program files to enhance management and control of these resources?

1206. Has the Execute permission been disabled on data folders if a user can copy into it?

1207. Are private files and public files kept in separate folders so that different access rights and permissions can be used for each?

1208. Are folder access privileges carefully limited to prevent users from having access to all subdirectories and files within that folder unless appropriate?

1209. Have the permissions on the Everybody group been changed so that this group does not have full access to all newly created folders?

1210. Are sensitive and "internal-only" files prohibited from being placed on a server which is connected to an untrusted network?

1211. Has the Auditing function been turned on for all password-protected and sensitive directories?

1212. Are files protected by the use of proper encryption?

Malicious Program Controls (Viruses, Trojan Horses, etc.)

1213. To help prevent viruses from infecting a program directory, are the permissions only set to Read and Execute?

1214. Are separate test systems used to determine if all new software programs are safe before they are installed on "production" network machines?

1215. To help curtail viruses, are Read Only permissions given to public file-sharing directories?

1216. Has a special virus scanning write only directory been created for users who need to put files on a server?

INFORMIX (Database Management System)

1217. Are the physical and logical logs stored on different disks, and do they minimize the chances of creating an I/O bottleneck?

1218. Are application executable and configuration files not in the same directory path as the application data files?

1219. In the file **$INFORMIXDIR/etc/sqlhosts**, do only user **informix** have read and write permissions on it, and all other users only have write permissions?

1220. Has a group been defined for database users which limits membership to a need-to-know?

1221. Does only the **informix** account have updated permissions to the control files?

1222. Is a single remote login account to the database application not permitted for multiple remote users?

1223. Is access to the database privileged account restricted?

1224. Is separation of user roles (e.g., SSO or SA) enforced, and if so, have different individuals been assigned to these roles?

1225. Are database users prohibited from being given the "grant with grant option" permission to database objects?

1226. Is the UNIX alias mechanism not used to treat two or more users as the same user within the database application?

1227. Are stored procedures and triggers set to prohibit users from inadvertently accelerating general user permissions?

1228. Is password and/or network encryption used to access all remote servers?

1229. Are the audit files owned by **root** with an Informix **group ID** and file permissions set to **600**?

1230. Are backups of the database performed, and are they performed on a regular basis?

1231. Does the database application use **/tmp directory** for temporary storage during execution to ensure that deletion of **/tmp directory** does not inadvertently delete online configuration files?

1232. Are any available secure database options used during database creation?

1233. Are the correct security options included in database startup and shutdown scripts?

ORACLE (Database Management System)

1234. Are the control and data files, the redo log files, and the audit files stored on separate drives?

1235. Is the **Oracle** account the owner of the **ORACLE_HOME** directory?

1236. Does only the **Oracle** account have update permissions for the control files?

1237. In the **/etc/groups** file, has a group been created for Oracle users?

1238. Is database client password encryption configured and enabled?

1239. Are the **UserIDs** and passwords not permitted to be hard coded in database links?

1240. In the table **DBA_DB_LINKS**, are none of the rows permitted to have the value PUBLIC in the username column?

1241. Is the operating system **I&A** not used for database **I&A**?

1242. Have the permissions on the database system tables been set to prohibit modification?

1243. Is access to the database privileged account restricted?

1244. Have default passwords been changed on any installation application login accounts?

1245. Are permissions for privileged roles been set to prohibit modification from the database application defaults?

1246. Do guest application login accounts exist?

1247. Is access to database objects prohibited to PUBLIC users?

1248. Are stored procedures and triggers set to prohibit users from inadvertently accelerating general user permissions?

1249. Are application level audits generated for the vendor-recommended events and to ensure the tractability of user action?

1250. Is the database **redo** log configured correctly, and is it being used to ensure recovery from failures?

1251. Are important database configuration files and logs being mirrored?

1252. Are application-provided utilities configured appropriately, and are they run on a regular basis?

SYBASE (Database Management System)

1253. Are application provided utilities, such as the **Database Consistency Check**, configured appropriately and run on a regular basis?

1254. Is the database restored from a previous backup and is the data verified to be accurate?

1255. Is no one able to change permissions on systems tables, with the exception of privileged users?

1256. Are application files, configuration files, and data files stored on the recommended drives or devices?

1257. Are logs assigned to separate devices to protect the system and audit databases when the system crashes?

1258. Are the **Sybase** executable and configuration files prohibited from being stored in the same directory path as the data files?

1259. Is the application directory owned by the proper user, and are the file permissions set correctly?

1260. Is the user **Sybase** the owner of the **$SYBASE** directory?

1261. Does only the **Sybase** account have update permissions to the control files?

1262. Has a group been defined for database users, and are they the only authorized users members of this group?

1263. Does the **Sybase** group only have read/execute permissions to all files under the **$SYBASE** directory?

1264. Is a single remote login account to the database application used for multiple remote users (since audit actions can only be traced to the local log-in)?

1265. Have permissions on the database system tables not been modified?

1266. Has access to the system tables in the master database been restricted to the system administrator account?

1267. Have default passwords been changed on any installation application login accounts?

1268. Have the installation application login accounts been deleted or locked after installation?

1269. Are users prohibited from setting the default database to the master database?

1270. Is separation of user roles (e.g., SSO or SA) enforced, and if so, have different individuals been assigned to these roles?

1271. Does only the system administrator have the privileges to create databases?

1272. Are SSOs only given permissions to access objects that are absolutely necessary to perform SSO functions?

1273. Are users prohibited to use **suid -1** and **uid 2**, which corresponds to a guest account (no guest accounts permitted)?

1274. Is an alias mechanism not permitted to be used to treat two or more users as the same user within the database application?

1275. Does audit ensure that the application asks for the correct role to be turned on for a user of a specific application?

1276. Are backups of the database performed, and are they performed on a regular basis?

1277. Are important database configuration files and logs being mirrored?

1278. Are any database auditing features, if they exist, configured and operational?

1279. Can only authorized users modify database audit features?

BACKUP CAPABILITIES

1280. Are all UPS (uninterruptible power supply) systems routinely checked and tested to ensure that they are functioning properly?

1281. To protect against data loss, has disk mirroring or duplexing function in Disk Manager been enabled for all drives?

1282. Have Backup Operator group members been carefully screened, and are these privileges absolutely necessary to each one's job function?

1283. Do Backup Operators have a special account only for performing backups which is separate from there normal user account?

1284. Are Backup Operator account activities logged and monitored?

1285. Are Backup Operator accounts restricted to specific systems and times?

1286. Are Backup Operator account passwords and names frequently changed?

WWW SECURITY

1287. Is Web server access limited to a designated area on the host?

1288. Have the Group ID and **UserID** been set to "Nobody" on the server?

1289. Has client access been limited to a specific directory, which has document **root** mapped to it?

1290. Has the director index been turned off unless there is an absolute need for it?

1291. Does access to sensitive document areas require the use of a password and a secure protocol such as **https**, **shhtp**, **PCT**, **SSL**, etc.?

1292. Is access to sensitive information areas controlled by a list of authorized hosts?

1293. Has the server's encryption key database and associated password file permission been set to **600** to protect this highly sensitive information?

1294. Are files such as key file, configuration file, etc., which may contain passwords, properly secured on the server?

1295. Has the script language interpreter been placed elsewhere other than the **cgi bin** since this creates a significant security vulnerability?

1296. Has user input been prevented from directly accessing the interpreter?

1297. Have unnecessary and unused network services and applications on the server been disabled?

1298. Have enabled network services and applications been properly secured, and are they actively monitored?

1299. Is the server's file system routinely checked to ensure that no unauthorized files can be accessed?

1300. If possible, is a separate **DNS** server running externally with only the minimum amount of information on it, since this helps prevent outsiders from doing **DNS** look-ups on the internal network?

1301. If possible, is the server run **chrooted**?

1302. Have only the essential and minimum amount of files been placed in **cgi-bin**? (Keep **csh**, **tcsh**, **perl**, **tclsh**, or other shell-like files out of this directory.)

1303. Has the web server been placed on an isolated machine so that it does not present a direct security risk to other critical departmental servers?

1304. Are the server's access and error logs continually monitored and audited?

1305. Have other mail programs, other than **Sendmail**, been considered as a more secure solution if appropriate?

1306. Have unnecessary and extra services been removed from the server (such as **ftp**, **nfs**, **telnet**, etc.)?

1307. If possible, is **NIS** being run so that **/etc/passwd/** is less useful?

1308. Are shadow passwords being used?

1309. Is **S/Key** being run for **root**?

1310. If used, are **/etc/passwd/** passwords on the Web server different from the ones used on other systems?

1311. Is the latest version of **http** being used so that the persistent and digest authentication functions can be implemented?

1312. Is the browser being used able to control active content when downloading, enabling, and during execution to help reduce **Active X** and **Java** security vulnerabilities?

Firewalls

1313. What kind of firewall is currently deployed (i.e., Packet Filtering, Application Proxy, Circuit Level, or Stateful Inspection)? If more than one, please note them, their type, and their location on the network.

1314. Is there a firewall deployed at the organization's Internet gateway?

1315. Does the firewall have the ability to perform Network Address Translation (**NAT**)?

1316. Has the firewall's Operating System (OS) been secured (usually done by the vendor)?

1317. Other than being improperly configured, has there been an incident where the firewall was breached?

1318. If so, did the breach originate from an external or internal source?

1319. Have corrective countermeasures been implemented to prevent any similar incidents?

1320. Does the firewall have a policy of blocking everything except for what is specifically allowed?

1321. Does the firewall block all connections if a power failure occurs?

1322. Does the firewall have the ability to block or allow attachments?

1323. Is **TCP wrapper** utilized by the firewall?

1324. Is **SOCKS** utilized by the firewall?

1325. Does the firewall help prevent applications from providing information about the network?

1326. Can dial-up connections be supported by the firewall?

1327. If so, can these dial-up sessions be encrypted by the firewall?

1328. Can the firewall be restored or replaced within one hour in the event of a major failure?

1329. If multiple firewalls are used, can they load balance among themselves to optimize their throughput?

1330. Does the firewall produce reports on network traffic and activities as well as attempted security violations?

1331. Are most of the firewall reports oriented toward exception transactions (security violations) rather than reporting on all transactions?

1332. Can the dial-up and Internet connections be authenticated by the firewall?

1333. Does the firewall support "challenge-response" authentication?

1334. Does the firewall support other authentication mechanisms such as RSA Public-Private Key Encryption, Kerberos, SecurID, S/Key, Digital Pathways, etc.?

1335. Does the firewall support firewall to firewall DES (56-bit or stronger) encryption?

1336. Can the firewall encrypt the connection between itself and a user, especially if the user is remote?

1337. Does the firewall use an automatic encryption key distribution and management system?

1338. Does the encryption key management system change the session keys at least once a day? (The more sensitive the information, the more frequently session keys should be changed.)

1339. Has the firewall had all compilers taken out of its system?

Intrusion Detection

1340. Is a real-time intrusion detection and response system being utilized?

1341. Are security log files from firewalls, routers, and servers (such as **SYSLOG**) continually monitored for possible attacks?

1342. To prevent attackers from modifying the security logs, are the logs sent to other append-only media?

1343. Are the **/bin/time** or **/bin/sh** directories (and other directories) routinely checked for **setuid** and **setgid** files which may have been left by attackers to give them **root** access?

1344. Are the system binaries routinely examined to determine if they have been altered? (Compare against a known good set to determine if any unauthorized changes have been made.)

1345. Have programs like MD5 and Tripwire been implemented to maintain the integrity of system binaries and other files?

1346. If so, is the output of these types of programs secured to maintain its integrity perhaps by using a cryptographic program such as Pretty Good Privacy (PGP) to sign it?

1347. Is the network routinely scanned for the presence of unauthorized network traffic analyzers or "sniffers?"

1348. Has the world-writable function been disabled on all jobs and job files running by "**at**" and "**cron**"?

1349. Is the **/etc/inted.conf** routinely examined to ensure that none of its programs have been substituted with a Trojan Horse program or modified to allow any attackers a backdoor?

1350. Is the **/etc/inted.conf** routinely examined to ensure that any entries, such as **/bin/csh** or **/bin/sh**, don't allow a shell program to execute to allow any attackers a backdoor?

1351. Is the **/etc/inted.conf** routinely examined to ensure that any previous disabled services have not been enabled to allow any attackers a backdoor?

1352. Is the **/etc/passwd** file routinely examined to ensure that any existing accounts have not had **UID** changes or accounts modified or created with no passwords?

1353. Have **.rhosts**, **/etc/hosts.lpd**, and **/etc/hosts.equiv** files been thoroughly examined for unauthorized nonlocal hosts and '+' entries?

1354. Are **.rhosts**, **/etc/hosts.lpd**, and **/etc/hosts.equiv** files not world writable?

1355. Is the file system frequently checked for unauthorized hidden files or unusual files?

TELECOMMUTING

1356. Are any users telecommuting or remotely accessing the network while traveling?

1357. If so, are there policies and procedures covering remote access?

1358. Do these policies define the method of connectivity and access control methods permitted?

1359. Has the physical security of the remote equipment been taken into consideration?

1360. Have externally accessible internal network resources been carefully selected and limited to an absolute minimum?

1361. Have specific policies and procedures been developed for the remote access of any proprietary information?

1362. Are there different levels of sensitive information classification, and can each be accessed remotely?

1363. Is there a form confirming that remote access users must sign affirming they will follow all established policies and procedures?

1364. Is there a policy which defines what is and what is not appropriate use of remote access equipment and resources?

1365. Is each remote user required to have a unique **UserID** and password?

1366. Is each remote user required to be properly authenticated before they gain any level of access to network resources?

1367. Are there authentication mechanisms other than **UserIDs** and passwords implemented, such as SecurID, Smart Cards, RSA, etc.?

1368. If a remote user's session becomes inactive, is there an automatic lockout or timeout feature?

1369. Are remote users only given three login attempts?

1370. Are all log-ins and other remote user activities monitored and logged for unauthorized events or actions?

1371. Has a log-in banner been implemented on your system so it appears to remote users before they log-in or immediately after they have logged-on?

1372. Does the log-in banner identify who owns the system?

1373. Does the log-in banner identify who is authorized to use the system?

1374. Does the log-in banner state that unauthorized use of the system is illegal and that it violates the system owner's policy?

1375. Does the login banner inform remote users that their activity is being monitored, and what will be done with this gathered information?

1376. Are remote users prohibited from uploading unauthorized software programs and files to internal network resources, since they may contain viruses and Trojan Horses?

1377. Are all remote user sessions authenticated and encrypted?

1378. Does the remote access system use an automated Public/Private key encryption system (such as RSA) to authenticate remote users?

1379. Does the remote access system use an automated symmetrical key encryption system for data privacy (encrypt all traffic) during all sessions?

1380. Is this automated symmetrical key encryption system using **IP** tunneling/encapsulation, which supports DES (56-bit), Triple DES (112-bit), and/or IDEA (128-bit)?

1381. Does the remote access system use SHA or MD5 for maintaining data integrity (cryptographic checksums)?

1382. Does the remote access system support multiple protocols such as IP, IPX, and NetBEUI?

1383. How many remote users does each remote access system support?

1384. Does the remote access system support company dial-in, ISP dial-in, or direct-connect PPP access methods?

1385. Other than establishing a PPP session and entering in a **UserID** and password, are remote users not required to perform any additional actions since the remote access system provides security in unobtrusive manner?

1386. Is the remote access system centrally managed and monitored?

TELECOMMUNICATIONS SECURITY

PBX Systems

1387. Is the PBX system located in physically secure area?

1388. Have all of the PBX vendor's default passwords and access methods been changed or disabled?

1389. If possible, have all maintenance and remote system administration ports been disabled?

1390. If maintenance or remote administration ports are used, do they require a maximum length passcode?

1391. Are all PBX system administration passwords changed frequently (every 30 days)?

1392. Does each system administrator have a unique **User ID** and password?

1393. Are all modems, voice mail systems, direct inward dialing features, Tie trunks, and tandem network services carefully secured from unauthorized access?

1394. Does the PBX system produce an audit trail for unauthorized entry attempts?

1395. Does the PBX system produce an audit trail for unauthorized calling patterns?

1396. Is there an audit trail of any changes made to the PBX system and who performed them?

1397. Are PBX system administration rights kept to a minimal number of individuals?

1398. Is the PBX system set to block or severely limit access after hours?

1399. Are all toll calls blocked or severely limited after hours?

1400. Are call detail reports carefully reviewed for patterns of teleabuse?

1401. Does the system generate exception reports for potential fraud patterns?

1402. Does the system block all 900-type calls and send an entry to the exception report?

Teleabuse Indicators

1403. Is the system continually busy?

1404. Are there any unexplained increases in international calling?

1405. Are there any unexplained or erratic calling patterns?

1406. Are there any unexplained increases in WATS and 800-number calling?

1407. Are there any unexplained increases in crank calls or wrong number calls?

1408. Are there any unexplained increases in after-hours calling?

1409. Does the time on hold seem to be unusually long for the system?

1410. Are calls prevented from being made inside voice mail?

1411. Is the system's call forwarding feature only limited to local users?

Direct Inward System Access (DISA)

1412. If it is necessary to use, is DISA access kept to an absolute minimum?

1413. Have the PBX vendor's default DISA passcodes been changed?

1414. Is each DISA user given a unique passcode?

1415. Is each DISA passcode a minimum of eight digits in length?

1416. Are DISA passcodes in random, not sequential, order?

1417. Does the PBX system silence when answering DISA calls?

1418. Are DISA phone numbers kept confidential?

1419. Are DISA phone numbers changed on a regular basis?

1420. Is DISA access blocked or severely limited after hours?

1421. Is DISA access blocked or severely limited to international calls?

Voice Mail

1422. Have all vendor default accounts been disabled?

1423. Have all vendor default system administration passcodes been changed?

1424. Are the system administrators aware of (and using) all security and audit functionality of the voice mail system?

1425. If possible, has DISA been blocked from voice mail?

1426. Have all collect call options been disabled or blocked?

1427. Has the option to transfer to other extensions been blocked from voice mail?

1428. Is each voice mail passcode a minimum of eight digits in length?

1429. Are voice mail passcodes in random, not sequential, order?

1430. Are voice mail passcodes changed every 30 days?

1431. If applicable, is the voice mailbox main access telephone number periodically changed?

1432. Are unused voice mailboxes disabled on the system (inactive or default accounts should be disabled)?

1433. Are voice mail users only allowed to attempt four consecutive log-ins?

1434. Are all unsuccessful log-in attempts alarmed and logged?

1435. Is the voice mail system routinely reviewed every 30 days for all inactive accounts, especially those which may belong to employees who have recently left the company?

1436. Are voice mailboxes monitored for any unusual activity or patterns?

Calling Cards

1437. Are calling cards used instead of offering DISA?

1438. Is the financial liability of each calling card limited to $50 if they are lost or stolen?

1439. Are employees trained on the proper and secure use of their calling cards?

1440. Are employees instructed to never give out their calling card and/or passcode to anyone?

1441. Are employees instructed to block the view of the telephone keypad when they are entering the calling card numbers to help prevent "shoulder surfing?"

1442. Are employees warned not to loudly speak their calling card numbers in public places?

1443. Are all calling card accounts regularly reviewed for irregular calling patterns?

1444. Are calling cards prevented from dialing internationally unless absolutely necessary on a per employee basis?

1445. Are calling card blocked from accessing 900 numbers and other types of call charge services?

1446. Are inactive calling card accounts disabled?

Cellular Telephones

1447. Are all cellular phones using digital encryption?

1448. Are cellular phone records examined every month to ensure that cloning is not occurring?

1449. Are all cellular phone users warned about cloning and how to look for suspicious patterns on the monthly call detail report?

1450. Are cellular phones blocked from calling 900 numbers and other types of call charge services?

1451. Are users instructed to keep the phones with them at all times (not left lying on a seat of their vehicle)?

1452. Are users instructed to keep their phones keypad locked with a passcode unless in use?

1453. If practical, can users keep their phones turned off unless they are making a call (this will reduce the likelihood of cloning.)

1454. If a user's phone is lost or stolen, are they instructed to immediately call their cellular provider and suspend their service?

1455. Are users instructed to never loan their phone to other individuals?

1456. Are users limited to a set dollar amount of cellular service they can use without an additional approval to limit the amount of financial risk exposure?

FINANCIAL SYSTEMS SECURITY

FINANCIAL — PAYROLL

1457. What method of employee payroll timekeeping is being used?

1458. If time cards are being used, are they frequently checked for accuracy and irregularities?

1459. Is a time clock used to document an employee's shift starting and finishing times?

1460. Is this time clock regularly checked to verify its accuracy?

1461. Do any changes made on the time card have to be approved by management?

1462. Is there a periodic audit during the shift of time cards and the personnel physically present to prevent someone from clocking in or out for a fellow employee?

1463. Once time cards are used, are they then voided to prevent unauthorized use?

1464. Are blank time cards or worksheets secured to prevent unauthorized use?

1465. Is each time card or worksheet marked with an individual serial number?

1466. Is an inventory maintained of all time cards or worksheets which is regularly audited?

1467. Is there an independent system used at irregular intervals for verifying accuracy of time and pay amount?

1468. Do employees sign a receipt voucher if cash disbursement is used for payroll?

1469. Are blank receipt vouchers secured to prevent unauthorized use?

1470. Is each receipt voucher marked with individual serial numbers?

1471. Is an inventory kept on all receipt vouchers, which is regularly audited?

1472. To facilitate an audit trail, are employees paid with checks?

1473. Is there a clear policy indicating that payroll disbursement will not be taken from current cash receipts?

1474. Are payroll preparation and the bank statement reconciliation of the payroll account handled by two separate individuals?

1475. Is the last step of payroll disbursement acquiring an authorizing signature? (This may preclude the inappropriate issuance of blank payroll checks.)

1476. If possible, are all paychecks physically given only to the employee listed on the check or automatically deposited in his or her checking account?

1477. Is the physical payroll check distribution process rotated among several different management individuals who are not directly involved in the payroll function?

1478. Are the employee payroll lists and time cards frequently compared to actual employee head count to dissuade workplace absentee fraud?

1479. Is each employee job payment code compared to his or her actual position and duties?

1480. Are payroll department employees periodically required to switch duties with others in payroll?

1481. Are job pay codes, changes in the rate of pay, hiring, termination, position advancements or reassignments, vacation time, sick leave, commissions, and bonuses approved by the appropriate management and kept current?

1482. Are hiring and termination approved by the appropriate management and kept current?

1483. Are position advancements and reassignments approved by the appropriate management and kept current?

1484. Are vacation time and sick leave approved by the appropriate management and kept current?

1485. Are commissions and bonuses approved by the appropriate management and kept current?

1486. Have all employees' tax withholdings been audited to determine if they have been deducted at the correct level and not manipulated by payroll personnel?

1487. Does the total of all employees' tax withholdings match the total tax liability owed to the government?

1488. Are changes in payroll policies and procedures approved by senior level management prior to implementation?

1489. Are all payroll policies and procedures periodically reviewed for legal compliance and operational efficiency?

1490. Are job pay codes and changes in the rate of pay approved by the appropriate management and kept current?

1491. Are administrators responsible for investigating unusual increases in labor costs and the number of employees?

1492. Is an investigation implemented when irregular or unclaimed payroll disbursements are discovered to safeguard against payroll padding?

1493. Is there an audit of all voided payroll checks?

1494. Who conducts this audit?

1495. Upon termination of an employee, is the personnel department responsible for contacting payroll services?

1496. Is the treasurer responsible for redepositing all unclaimed payroll disbursements?

1497. For limited disbursements, is there a petty cash fund available?

1498. Has a strict policy been implemented to prevent commingling with all other accounts?

FINANCIAL — ACCOUNTS RECEIVABLE

1499. Who is responsible for checking the work of the accounts receivable personnel?

1500. Do employees not involved in adjustments and credit memos handle accounts receivable record changes?

1501. Are vouchers for credit and any sales receipts serial numbered and completely accounted for?

1502. Are customers required to sign receipts of sale, even if the sales are cash or credit transactions?

1503. Does a customer get approval from the properly authorized person before any credit or extensions are granted?

1504. Are signed credit vouchers required before merchandise is returned?

1505. Are monthly statements mailed to customers showing the current status of their accounts?

1506. What are the procedures for customers to handle any discrepancies in their accounts?

1507. Are customer accounts randomly audited for accuracy?

1508. Are there delinquent credit accounts over 30, 60, and 90 days?

1509. Have accounts over 60 days been turned over for collection?

1510. List any exceptions and why these accounts have not been forwarded for collection.

FINANCIAL — ACCOUNTS PAYABLE

1511. Are accounts payable credits and credits for returned products properly recorded?

1512. Do adjustments to the accounts payable ledger have a valid paperwork audit trail?

1513. Are there any discrepancies in vendor invoices and quantities received?

1514. Are there any discrepancies in vendor invoiced cost versus the actual cost on purchase orders?

1515. Are there any discrepancies in the quality of vendor invoiced items versus what was ordered?

1516. Are all payments to accounts payable for the correct accounting period?

1517. Are accounts payable employees who are found to frequently change their billing methods continuously audited for possible fraud?

1518. Are all accounts payable balances carefully checked to ensure a debit balance doesn't mistakenly get issued a payment?

1519. Are all open balance accounts payable aged in 30-day segments?

1520. Does the general ledger control balance match the total of the individual accounts payable?

1521. Are accounts payable routinely audited by someone outside of the department?

1522. Have there been any incidents of suspected vendor fraud?

1523. If there have been incidents, was there any evidence of collusion with an employee?

1524. Are gratuities prohibited to individual employees, especially those with purchasing and accounts payable responsibilities?

1525. If gratuities are permitted, does a company-designated executive first review them?

1526. If gratuities are permitted, is there a dollar amount limit set, with the prohibition of cash awards?

1527. Are all changes in purchasing policies and procedures approved by senior management?

FINANCIAL — PURCHASING

1528. Have all purchasing functions been centralized and delegated to one person or tightly controlled group who may increase the probability of fraud?

1529. Does a copy of the invoice being paid accompany the check for payment?

1530. Are purchase orders and invoices cross-referenced before payment is authorized?

1531. Are triplicate format purchase order forms used?

1532. Is each sheet of the triplicate form a different color?

1533. Is each sheet of the triplicate form marked for its intended routing purposes?

1534. Is there a system established for authenticating the originality of sales and purchase orders to detect any deviations or alterations?

1535. Are check numbers and dates noted on all paid invoices?

1536. Do both the invoice and the check for payment bear proper authorization?

1537. Is there a system for keeping track of all returned purchases?

1538. Are returned purchases processed by someone other than the person(s) who conducted the original purchase?

1539. Is all vendor application information thoroughly researched and verified before the vendor is placed on the approved list?

1540. Are all preapproved vendors kept on a list, which is regularly updated?

1541. Are back orders canceled after a designated period (30, 60, 90 days)?

1542. Are unfilled purchase orders left open as back orders?

1543. Are all purchase order forms sequentially numbered, inventoried, and audited?

1544. When certain inventory items and supplies are depleted to a designated level, are they automatically reordered?

1545. Do purchases over a designated dollar amount require senior management approval?

1546. Are two signatures required on all purchase orders?

1547. Are all products received before a payment is authorized?

1548. Have all available accounts payable discount schedules been taken?

1549. Are the dates of all payments consistent with discounts received?

FINANCIAL —CASH DISBURSEMENT

1550. Are cash disbursements separated from cash receipts?

1551. Are cash disbursements recorded on a special form or program and reconciled daily?

1552. Are cash disbursement forms prenumbered from an audited inventory?

1553. Is an explanation section for the cash disbursement included on the form?

1554. Are these forms required to be complete and detailed?

1555. If cash disbursements (refunds or paid-outs) are made from cash received (receipts), do they require an authorization from management?

1556. Are any variances in cash funds immediately investigated?

1557. Who is responsible for scrutinizing the entries for bad debts, returns, allowances, discounts, and all other noncash items?

1558. When a customer is listed as an uncorrectable account, are steps taken to determine whether or not he or she actually exists?

1559. Is there a procedure for tracking overdue accounts?

1560. Does an annual audit verify customers' unpaid balances?

1561. Who is responsible for this audit?

1562. Is there an audit trail for every adjustment made to customers' accounts?

1563. How often are unannounced audits made?

1564. Does an outside-certified public accounting firm perform the annual audit?

1565. Is a random and unannounced position switching schedule used for employees working with collection and credit accounts?

1566. Are cash ledger entries performed by someone other than the individual in charge of tabulating cash on hand or bank reconciliation?

FINANCIAL — PETTY CASH

1567. Is petty cash kept at the facility?

1568. Are petty cash funds kept to a minimum?

1569. What is the amount of petty cash kept?

1570. Are these funds subject to random and frequent audits?

1571. Where is the petty cash located when in use?

1572. Where is the petty cash secured when not in use (safe, locked desk, or file cabinet)?

1573. When in use, is the petty cash fund kept in a locking metal box or drawer?

1574. When in use, is the petty cash fund kept in a cash register?

1575. When in use, is the petty cash fund storage device physically secured to a solid structure to prevent someone from easily taking the storage device?

1576. Is the day storage (in-use) area under constant employee supervision?

1577. Is this the same area for night storage?

1578. Is the night storage area accessible only to management?

1579. Are day and night storage areas secured?

1580. Are locks in petty cash storage areas replaced when keys are lost, missing, or stolen?

1581. How many employees handle the petty cash funds?

1582. Is the number of employees authorized to handle petty cash transactions kept to an absolute minimum?

1583. Are the signatures of the person receiving and the person authorizing funds required on a voucher?

1584. Is a completed voucher required before any disbursements are made from the petty cash fund?

1585. Are all petty cash vouchers accompanied with the original invoices?

1586. Is the amount of the voucher both written out and designated with numerals?

1587. Is permanent-type ink used on all vouchers?

1588. Are numerically sequenced petty cash fund vouchers used?

1589. Does the amount of the check issued to replenish petty cash equal the amount of related paid-out vouchers?

1590. Are blank checks also located with petty cash?

1591. Are these checks prohibited from being presigned?

1592. Is someone other than the person responsible for petty cash verifying and canceling vouchers when the fund is replenished?

1593. Is there a system for checking paid invoices to ensure that they have not been tampered with?

1594. Who is responsible for performing unannounced and unexpected audits of the petty cash fund?

1595. Is the level of fund usage routinely monitored to justify the amount kept in petty cash to avoid excessive funds?

1596. Are personal cash advances or IOUs prohibited from being disbursed from petty cash?

1597. Is cashing personal checks prohibited from the petty cash fund?

1598. Are cash donations prohibited from being disbursed from petty cash?

1599. Are transactions, which can be or normally are, paid with a company check prohibited from being disbursed from petty cash?

1600. Are all discrepancies that occur in the petty cash fund immediately investigated by security?

1601. Is the petty cash fund subject to random and frequent audits?

1602. Is the petty cash fund balanced daily?

1603. Is all petty cash fund activity tracked and recorded in a daily journal book?

1604. Does an individual who is not directly involved in petty cash fund disbursements reconcile this daily journal?

1605. Does the daily journal amount agree with the general ledger amounts for petty cash?

1606. Is management approval required for petty cash fund vouchers over a designated dollar amount?

1607. Is senior management approval required for any changes in petty cash fund policies and procedures?

FINANCIAL — COMPANY CHECKS

1608. When a cash disbursement is made, are the appropriate general ledger accounts credited independently?

1609. Except for payouts from petty cash, are all disbursements paid by checks?

1610. Are two signatures required for check issuance?

1611. Are the blank checks issued to a location or department tracked and accounted for by using an inventory log?

1612. Is the inventory of blank checks routinely audited to ensure that none are missing, especially toward the bottom of the stack?

1613. Do checks contain a printed statement on their face which voids them if they are not cashed after a certain time period from issuance?

1614. Are all of the company's payments required to be made by check?

1615. If a signature stamp is used, is it kept in the safe in an unmarked bag or envelope?

1616. Are checks signed by a different person from the one issuing a voucher?

1617. Are handwritten checks, rather than being check-written and typed, prohibited from being issued?

1618. Are checks only sent to an approved list of vendors?

1619. Is there a serial numbering system used for all checks?

1620. Are checks with errors voided and new ones issued in their place?

FINANCIAL — EXPENSE CHECKS

1621. Has any person outside of management signed expense checks?

1622. Have expense checks been used for any unauthorized purchases?

1623. Are any expense checks missing?

1624. Are voided check copies sent to the Sales Audit department?

1625. Are expense checks issued in sequence?

1626. Are all personal expense accounts carefully monitored and reviewed?

1627. Is a valid receipt required for all reimbursed expenses?

1628. Is there a limit placed on expenses such as hotel, meals, and entertainment?

1629. Is there a business activity log or reason section on the expense summary form for providing details and explanation for expenses?

1630. Is the activity log verified as accurate? (For example, a salesperson claims to have visited a business. However, when the business was called, they said he had not called on them during the period listed by the salesperson.)

1631. Is this same level of scrutiny especially applied to all entertainment expenses? (For example, a salesperson listed several meals as client expenses. However, an inquiry to the listed clients revealed they were not taken to dinner as listed. An investigation discovered that the salesperson was taking friends and family out to dinner.)

1632. Do accounting or security personnel routinely verify expenses?

1633. Are expenses such as mileage carefully reviewed to verify what is listed is correct (e.g., a trip is claimed as 500 miles when it is 250 miles)?

1634. Is the fuel mileage rating known for each individual claiming a fuel expense?

1635. Are all receipts closely examined to detect any alteration of numbers?

1636. Are some nonreceipted items such as parking and toll fees verified to be accurate as claimed?

1637. Are airline tickets or any travel expenses audited for accuracy? (For example, a vice-president listed an airline ticket as first class. However, he had returned the first-class ticket, bought a coach ticket, and pocketed the difference.)

1638. Are all company-related expenses required to be charged on company-issued, not personal, credit cards?

1639. Are all fuel and related expenses required to be purchased on company-issued gasoline credit cards?

1640. Are all airline and related travel arrangements to be coordinated and authorized through the company's travel service?

FINANCIAL — INVENTORIES

1641. Are inventories physically counted?

1642. If not, what method is used to determine inventory levels?

1643. Do the physical inventories agree with the adjusted general control accounts?

1644. Were consignments in or out of inventory levels reflected when taking a physical count?

1645. What is the basis for determining the market value of the inventory?

1646. Are labor, material, and overhead expenses included in inventory costs?

1647. Does the inventory reflect write-downs for cost over the net realizable value or obsolescence?

1648. Do the cutoff dates for returned goods, purchases, and goods in transit correspond with the current inventory determination period?

1649. What are the procedures for intercompany and interdepartmental sales of inventories or transfers?

1650. Who is responsible for these transactions?

1651. What is the amount of shrinkage in the inventory?

1652. Is this figure above the normal level, given normal business operations?

1653. Is a significant amount of shrinkage attributable to theft?

1654. Is this theft external or internal?

1655. Is the business involved in its own retail operations?

1656. Does the business ship to wholesalers and/or retailers?

1657. How much shrinkage occurs in the retail environment?

1658. How much shrinkage occurs before inventory arrives at the retail or customer locations?

1659. Who is responsible for controlling the receiving and shipping of inventory?

1660. Do two separate individuals perform these functions?

1661. Are the receiving and shipping areas physically separated?

1662. Are random audits of the inventory conducted by someone outside the department?

1663. Are there surveillance cameras located in the inventory area?

1664. Who investigates discrepancies or theft related to the inventory?

1665. Are high value items stored in a special area with additional physical security considerations?

1666. Are all employees carefully screened through background checks and honesty tests?

1667. Is there an incentive or bonus program available for inventory or warehouse employees if they reduce shrinkage in the inventory?

1668. Are there any special factors or events that affect the current valuation of the inventory?

RETAIL SECURITY

RETAIL — CASH REGISTER ACCOUNTABILITY

1669. Do sales staff use only one register per clerk?

1670. Is each cashier given an identification number so that each transaction can be tracked to the responsible cashier?

1671. If a salesperson or manager is using the same register as a cashier, are they also given individual identification numbers?

1672. Does a responsible individual keep the cash register's currency storage to a minimum amount to make change?

1673. Is there a policy for having the registers cleared at least twice a day?

1674. Are registers randomly audited during a cashier's shift to balance actual funds versus recorded transactions?

1675. Are registers programmed not to operate if the drawer has been left open from the previous transaction?

1676. Are all types of transactions tendered through the register and then given receipts?

1677. Does management personnel have exclusive control over the register operation keys and post void procedures?

1678. When UPC scanning devices are used, are physical recounts of purchased merchandise done to verify accuracy of the cashier versus the register?

1679. Are sales invoices being used in numerical sequence?

1680. Are all invoices accounted for?

1681. What method is used to signify that all sales receipts reflect when a sale is made?

1682. Are over-the-counter receipts issued by cash register?

1683. Do the registers also use prenumbered receipts?

1684. Is there someone other than the sales staff responsible for the processing of duplicate receipts to maintain the system's integrity?

1685. Does the register provide for a duplicate copy inside the machine, which only management has access to, to prevent possible tampering from employees?

1686. Is there someone other than the sales staff responsible for verifying that the amount of cash turned in agrees with the duplicate receipts?

1687. Does someone other than the sales staff have the keys to control the locks and the functioning mode switches of each register?

1688. Were discounts found that were not recorded on the discount log sheet?

1689. Did a member of management initial the invoice authorizing the discount next to the reduced price of the item?

1690. If hand written tickets are used for over-the-counter receipts, are they prenumbered for accountability?

1691. How many copies of prenumbered receipts are made (duplicate or triplicate)?

1692. Does the amount of cash received agree with the receipts for each shift?

1693. Are all change requests done out of a change drawer or fund rather than out of other cashiers' registers?

1694. Are all "no sale" receipts accounted for and attached to a cashier's daily report?

1695. Is there a daily "Over and Short" log kept for each person and/or register?

1696. Are "Overs and Shorts" thoroughly investigated and monitored to determine whether each is either an error or theft?

1697. Are there any patterns in the occurrence of "Overs and Shorts," such as just before pay-day, just before Friday and Saturday nights, or just before an employee's days off?

1698. Are "Overs and Shorts" occurring in a "shotgun" fashion?

1699. Are there any recurring patterns in "Overs and Shorts?"

1700. Are all cashiers and salespeople periodically integrity (honesty) shopped with a unique product purchased with the exact amount to verify the transaction was rung up correctly?

1701. During the integrity shop, are receipts intentionally left behind to see if they show up as voids or refunds later on?

1702. Are all integrity shops thoroughly and completely documented with specific items, amounts, registers, times, conversations, and cashiers/salespeople in case of their use in theft prosecution?

1703. Are all employee names compared against credit card transactions?

1704. Is the sales staff responsible for each prenumbered receipt?

1705. Is a cashier's fund also counted and verified by management?

1706. Are proper security measures taken when a cashier has to turn in or pick up money from management?

1707. Do the cashiers have a register cash count form that has been completed?

1708. Pull the daily correction report from the cashiers. Are there any “no sale” transactions? If so, are the reasons for the “no sale” or corrections recorded on the receipt?

1709. Are the transaction numbers and totals listed?

1710. Check the correction report for post voids. Are the reasons for the voids stated on the back of the receipt?

1711. Are corrections such as total voids and post void transactions approved by management at the time of the transaction?

1712. Are there any total voids that have not been approved by the management?

1713. Are there any post void transactions that do not have the initials of the manager?

1714. Are total voids and post voids recorded on the daily correction record?

1715. Are there any personal items in the cashier's work station?

1716. Are the cashiers closing the drawers after each customer has been helped?

1717. Are cashiers examining all merchandise packaging to detect concealed items?

RETAIL — CREDIT CARDS

1718. Do credit cards receive the proper authorization by telephone or terminal?

1719. Do cashiers carefully compare the credit card bearer's identity with the name on the card?

1720. Are all credit cards checked carefully to ensure that they have not expired?

1721. Are all credit cards checked carefully to ensure that they are not stolen, altered, or counterfeited?

1722. Are all credit card receipts left behind by the customer immediately destroyed?

1723. Is the customer only allowed to sign the credit card voucher after an authorization code has been recorded on the charge slip and their identity been verified?

1724. Does the register system identify when a transaction involves a credit card?

   A. Does it record the card number and reference both numbers to each other for auditing purposes?

1725. Is all information concerning credit cards, especially account numbers, adequately safeguarded?

1726. Are blank vouchers and forms physically secured and accounted for by regular inventory and tracking of these items?

1727. Are refunds or credits only issued directly to the credit card account number rather than giving customer's cash back?

1728. Does the register system have a special report to monitor an excessive amount of credits to any one credit card account or one individual?

1729. If a customer refutes a purchase, are they made aware that credit card purchases conducted by telephone may be legal unless there is a signed voucher?

1730. Do cashiers and salespersons carefully compare all credit card transactions and slips for multiple transactions using the same card during one shift?

1731. To identify counterfeit Visa and MasterCard credit cards, are personnel trained to compare the first four to six digits of embossed account numbers on the card with the Bank Identification Number (BIN) directory which lists legitimate credit card issuing banks?

1732. Are personnel aware that some credit cards such as Visa also print the BIN numbers in smaller type (not raised) directly above the first four to six embossed numbers in the card's account number?

1733. Are personnel aware that some credit cards such as Visa use microprinting of the BIN number to form a frame around the Visa symbol on the face of the card?

1734. Have employees received specific training to recognize altered or counterfeited credit cards?

1735. If employees believe someone has given them an altered or counterfeited card, do they call the bank to verify that the card's and the presenter's information match?

1736. Does the company have a current list of fraudulent credit card information?

1737. Is this fraudulent list compared against any orders placed before they are shipped?

1738. Is there a list of contacts for credit card fraud centers for each type of card accepted by the company?

1739. Does the company bond all employees responsible for processing credit cards?

1740. Has the fraudulent credit card list been examined for any employee names?

1741. Are all suspected credit card frauds through the mail immediately turned over to a postal inspector?

1742. Does the company receive any business crime awareness newsletters or other publications that discuss current issues concerning fraud?

1743. Are credit card processing procedures periodically reviewed to ensure that they address any increased activity in, or new methods of, fraud?

RETAIL — CHECKS

1744. What is the company's check cashing policy?

1745. Are these policies checked by legal counsel for any type of discriminatory procedures?

1746. Are all checks required to be drawn on local banks?

1747. Is proper identification, usually state-issued driver's license or ID card, required of the person cashing the check?

1748. Are checks refused if they are dated in advance of the current day or appear to be possibly altered in any way?

1749. Do employees compare the signature on presented checks against the proper identification?

1750. Was the check written and signed in the presence of the employee?

1751. Do employees verify that the numerical amount and the written amount agree?

1752. Do employees accept only legible checks?

1753. Are checks accepted only for the exact amount of purchase?

1754. Are two-party checks accepted?

1755. Are payroll checks accepted?

1756. Are government checks accepted?

1757. Are money orders and travelers' checks accepted?

1758. Are only company and personal checks made payable to the business accepted?

1759. Are starter checks and low-numbered checks refused?

1760. Are checks, which exceed a specified limit, verified by calling the issuing bank?

1761. Are only checks written in ink accepted?

1762. Are there specific procedures governing the acceptance of each individual type of check?

1763. Are employees instructed not to accept a check when the person attempting to cash it is intoxicated or acts suspiciously?

1764. Does the company have copies of the complete federal, state, and municipal laws governing check processing and frauds?

1765. Does the company actively prosecute bad check writers and pursue reimbursement through civil recovery?

1766. What procedures are recommended by the company's bank for reimbursement of bad checks?

1767. Are credit bureaus, merchant associations, or check verification services used as a preventative measure?

RETAIL — BANK DEPOSITS

1768. Are cash entry records compared with the cash receipts list to verify each shift's accuracy?

1769. Are the separate functions of bookkeeper and cashier divided among separate individuals?

1770. What procedures have been implemented to ensure that neither the cashier nor the bookkeeper has access to one another's records?

1771. Have procedures been implemented to prevent the mail from being opened by a cashier or bookkeeper?

1772. Is there a duplicate record on numbered forms, signed by a fiscally responsible individual and cashier, kept to verify original documentation of cash receipts?

1773. Does the auditing department receive a copy of the cash receipts record while the cashier keeps the original?

1774. Is the bank deposit receipt returned to the cashier responsible for it?

1775. Do the daily deposits reflect the amount of receipts?

1776. Are procedures established for depositing checks within a day of receiving them?

1777. Have all received checks been stamped with "for deposit only"?

1778. How many slips are prepared for each bank deposit?

1779. Who checks the cash receipts record on a weekly basis?

1780. Does the bank mail copies of deposits to the accounting or auditing department?

1781. Is one person used for reconciling bank statements while another handles deposits or withdrawals?

1782. Does the amount of cash and checks received balance with the daily deposits?

1783. Is someone other than the head of the accounting department responsible for reconciling bank statements?

1784. Who is responsible for the examination of canceled checks to detect tampering, alterations, and proper endorsement?

1785. If any irregularities appear on bank statements, are duplicate copies obtained from the bank?

1786. Are stringent controls placed on the sale of miscellaneous items such as used equipment, scrap materials, and salvage?

1787. Has a procedural audit program been specifically developed to detect internal theft activities?

1788. Who is responsible for conducting the audits?

1789. Are invoices and a billing system utilized to eliminate the need for cash transactions?

1790. If not, are all transactions involving the exchange of cash recorded immediately and recorded on receipts?

1791. Are these receipts reconciled daily with the actual cash count?

1792. Is a cash register used to produce an audit trail in the form of a detailed journal tape?

1793. Is the cash received (receipts) deposited in the bank daily?

RETAIL — REFUNDS

1794. Check all store files for refund check payments. Are the payment copies kept on file?

1795. Is the check acceptance policy visible to customers at the register?

1796. Are all cashier's refunds periodically verified to determine if they are legitimate?

1797. Are all receipts carefully examined to determine if they are legitimate?

1798. Are all refunds documented with the properly completed forms?

1799. Have a random number of completed refund forms been verified to be legitimate by contacting the customers listed?

1800. Review the refund documents. Are the receipts attached? (If not, is there a notation as to the status of the original receipt?)

1801. Are the refund documents completely filled out with the customer signature, validation, employee name, and number, etc.?

1802. Are the refund documents kept in an orderly manner?

1803. Are driver's license numbers recorded on refunds without receipts?

1804. Do refunds in excess of a designated amount have a manager's approval?

1805. Are paid-outs accounted for and logged on paid-out control forms?

1806. Has a list of persons attempting fraudulent refunds been compiled and placed at every register and the refund counter?

RETAIL — STORE SAFE

1807. Are funds in the safe kept to an absolute minimum?

1808. Where is the safe kept?

1809. Is the safe UL (Underwriter Laboratories) listed for both fire and burglary resistance?

1810. Is the safe attached to a floor or wall structure?

1811. Is the safe combination restricted to a minimal number of employees and changed periodically?

1812. Is it mandatory that the safe lock be completely reset every time it is used rather than being left in "day-lock" mode?

1813. Is the safe equipped with a keypad combination lock that can record all activity by individual passcode numbers?

1814. Is the safe combination always changed immediately when an employee is terminated?

1815. Is the safe equipped with a time-lock feature?

1816. Is the safe always checked to ensure it is locked as part of the daily closing procedures?

RETAIL — RECEIVING

1817. Are all incoming shipments required to be documented immediately?

1818. Are purchase orders matched with the corresponding receiving documents?

1819. Are all adjustments made to purchase orders relayed to the receiving department so that arriving shipments will have an accurate count?

1820. Are there only certain hours when shipments can be received?

1821. Are bills of lading complete?

1822. Who is responsible for receiving materials?

1823. Does this person also prepare the documentation for receiving?

1824. Is this person separate from the purchasing department?

1825. Are all incoming shipments thoroughly inspected to ensure the quantity shipped matches the quantity received?

1826. What are the procedures for reporting and handling discrepancies?

1827. Are any discrepancies noted and identified before any receiving documentation is signed?

1828. Are incoming materials inspected for damage?

1829. Are all empty boxes immediately removed from the receiving areas?

1830. Is the high value storage area protected by additional intrusion detection equipment?

1831. Is this alarm system capable of being activated independently of the host building's system to provide security during regular operational hours?

1832. What are the procedures for reporting and returning damaged materials?

1833. Who is responsible for these return authorizations?

1834. Are there unusual entries in the UPS log (e.g., addressed to employees, individual people, etc.)?

1835. Are the following files retained?

A. Outside vendor orders

B. Special orders and rain checks

C. Defectives pending authorization

D. Defectives (Vendor Merchandise Returns — [VMR] claims) copies

E. Freight claims — pending inspection

F. Freight claims — copies

G. Bills of lading

1836. Review the "Receiving Completed" bin. Has paperwork been sent to the corporate office daily?

1837. Are the receiving documents kept in order and organized?

1838. Are all pedestrian and roll-up doors secured when a shipment is not being unloaded?

1839. Is there any unusual, hidden, or loose stock in the receiving area?

1840. Check the returns counter and the receiving area for defective merchandise. Are all defectives tagged with a completed defective sticker?

1841. Review the vendor merchandise return forms. Have the sales representatives signed for the merchandise when it was picked up?

1842. Is the defective merchandise area neat and well organized?

1843. Is all defective merchandise less than 90 days old?

1844. Is all defective merchandise inventory at, or below, the designated level?

1845. Are any discrepancies noted and identified before any receiving documentation is signed?

1846. Is there a form and procedures for dealing with damaged freight merchandise?

1847. Is there a form and procedures for handling freight shortages and overages?

1848. Is a copy of the sales order required to be attached before merchandise to be shipped leaves the warehouse?

1849. Is there a two-person, double-checking system for the accuracy of merchandise being shipped and where it is being sent?

1850. Who is responsible for preparing sales orders?

1851. Are the sales order forms consistently accounted for and prenumbered?

1852. Is the shipping area physically separate from the receiving area?

1853. If this is not possible, are deliveries and shipments arranged so they do not occur at or near the same times?

1854. Are the loading doors to the shipping and receiving areas closed when not loading or receiving a shipment?

1855. Are employees prohibited from parking in the immediate vicinity of these areas?

1856. Are all freight containers and vehicles locked when not in use?

1857. Are there any objects obstructing fire exits and lanes?

RETAIL — SHOPLIFTING

1858. Are employees trained to look out for the following methods of shoplifting?

1859. Have shoplifters been ticket switching one expensive piece of merchandise with a lesser-priced item?

1860. Are individuals going to remote/secluded areas or exits of the store to attempt to conceal items to leave the premises undetected?

1861. Are there individuals concealing merchandise in a baby stroller, carriage seat, or on a child's person?

1862. Is anyone using paper bags, umbrellas, large purses, briefcases, diaper bags, backpacks, and gym bags to conceal merchandise?

1863. Is anyone wearing baggy or out of season clothing capable of hiding multiple items?

1864. Is anyone wearing specially designed clothing with an expandable waistband or leg-bands to hold merchandise that is then worn under loose-fitting garments?

1865. Are individuals wearing overcoats customized with hidden pockets and/or hooks on the inside of the coat's lining?

1866. Have there been several shoplifters working as groups to distract employees while others shoplift?

1867. Has merchandise been placed by a store exit for the “grab and run” technique?

1868. Are employees trained to recognize shoplifters' characteristics?

1869. Are there individuals who look at numerous but dissimilar items or just don't know what they want to purchase?

1870. Are there individuals who “accidentally” wander into areas that are clearly marked (or intended) for employee use such as display cases or stock areas?

1871. Are there individuals who are in an unusual hurry to exit the store?

1872. Are there individuals who are paying more attention to where the store personnel are located than the merchandise they are supposedly selecting?

1873. Are there nervous individuals with rapid, scanning eye movements?

1874. Are there individuals who walk in a restricted or peculiar manner or have awkward body movements in an attempt to conceal stolen items?

1875. Are there individuals wearing out of season or baggy clothing which could conceal merchandise?

1876. Are there individuals who may create a scene to distract employees?

1877. Are there individuals who are just “waiting” for someone who is shopping?

1878. Are employees trained in the following methods of preventing shoplifting?

1879. Does the physical design of the store control access by limiting the number of entrances/exits, and keeping exits clear of merchandise items and displays that could obscure an employee's ability to watch an area?

1880. Are all areas and displays kept neat and clean?

1881. Are employees assigned to certain sections of the store?

1882. Does an assigned employee cover each area of the store during peak hours?

1883. Do price tags tear apart when attempts are made to remove them?

1884. Are items which are susceptible to ticket switching double tagged in a concealed area of the merchandise to help detect switching?

1885. Is an Electronic Article Surveillance (EAS) system in use?

1886. Do cashiers carefully inspect merchandise for tag switching or incorrect pricing?

1887. Do cashiers carefully inspect merchandise for additional items hidden inside legitimately purchased merchandise containers?

1888. Are purchased items placed in a bag and stapled shut with the receipt on the outside for every sale?

1889. Are all higher value items that are openly displayed secured to the display structure?

1890. Are small, high-value items placed in a locked display case with pick-up tickets? Is it visible to numerous employees?

1891. Are empty boxes and containers removed from display areas?

1892. Are inventories kept accurate and current to identify areas of shrinkage?

1893. Are plainclothes and/or uniformed security personnel utilized?

1894. Are all employees instructed to immediately greet all customers entering the store and in each employee's area?

1895. Does the store use convex mirrors, closed circuit television, and intrusion detection alarms?

1896. Are all shoplifters criminally prosecuted and processed for a civil demand claim?

RETAIL — POINT OF SALE (POS) SYSTEMS

1897. Is the company involved in a retail operation that utilizes a POS system?

1898. Does the system track a perpetual inventory?

1899. Does each sale, refund/return, and special-order transaction affect the perpetual inventory?

1900. Does the POS system track items by category and record their gross margins?

1901. Does the system track the following?

A. Performance of sales people

B. Individual cashier's performance

C. Exception transactions such as voids, refunds, and "no sales"

D. Overages and shortages by cashier or sales person

1902. Is there a special report that compiles all exception transactions by cashier/salesperson and tracks them over an extended period of time to detect potential internal theft problems?

1903. Are all profiled exception report trends investigated for potential internal theft?

1904. Are there specific levels of access and certain areas within the system?

1905. Are store level personnel, including management, prevented from having software programming changing password access?

1906. Do the following items require management intervention?

A. Changing prices

B. Changing costs

C. Changing perpetual inventory levels

D. Running daily register totals

E. Running any type of summary report

F. Changing quantities received

G. Finalizing purchase orders

H. Processing post voids

I. Processing refunds over a specific dollar amount

J. Accessing the system's training mode

1907. Are all employees, except for management, prohibited from making any system changes?

1908. Can management make system operational changes only after they have contacted the data center and obtained approval?

1909. Does the system record all detail of each transaction?

1910. If the cash drawer is not closed after a transaction, does the system not permit another transaction to be processed until the drawer has been properly closed?

1911. Is a register electronically locked if a cashier/salesperson has to leave the area?

1912. Does the register system only allow one user at a time to be signed on?

1913. Are cashiers/salespersons strictly assigned to one register?

1914. Is cash drawer key-issuance limited to management personnel?

1915. Is access for a terminated employee immediately removed from the system?

1916. Are all system documentation and software secured?

1917. Is system equipment physically secured and inventoried by serial number?

1918. Has the media storage area been adequately secured against casual theft attempts?

1919. If individuals other than data center personnel need access to materials, is there a system to document their activities and limit their access?

1920. Has a plan been developed to handle emergency situations and to notify the proper personnel?

1921. Have procedures been developed to monitor computer usage and activities?

1922. Who is responsible for monitoring system usage information?

1923. Have control logs been incorporated into the usage monitoring?

1924. Have all critical data been identified and safeguarded from theft and damage?

1925. Do individuals need personalized access codes to activate a system terminal?

1926. Are these codes changed periodically and tracked for their type of activity?

1927. Are employees warned never to give their codes to another person or to use anyone else's code?

1928. Are proprietary software programs protected?

1929. Are all operating systems and other software backed-up and protected from theft and damage?

1930. Is software documentation protected at the same level as actual software media?

1931. Are there established procedures for changing software and its protection during this process?

1932. Is all input processing carefully controlled and does it follow established standards?

1933. Are all generated reports secured and only released to those individuals who are authorized to receive the data?

1934. Is report-generating activity monitored and documented by management personnel?

1935. Is there an established resource scheduling system for the data processing center?

1936. Can the system be accessed by modem?

1937. Does the dial-in modem have a password security feature before system access is allowed?

1938. Does the modem dial back the user to verify their identity before they are allowed system access?

1939. After three unsuccessful logon attempts, does the modem automatically deny system access?

1940. Are modem passwords different than data system access codes?

1941. Are the modem passwords changed frequently and under a strictly limited distribution?

1942. Are modem password issuances, changes, and terminations thoroughly documented?

1943. Does the system administrative software document all log-in attempts, especially unsuccessful ones?

1944. Does the system administrative software also identify unsuccessful log-in attempts from within the system?

1945. Are all unsuccessful log-in attempts immediately investigated to determine their source?

1946. Does the system send an alarm message to data center personnel after an unsuccessful log-in attempt?

1947. Is the security department immediately notified when unauthorized access attempts are detected?

1948. Is there restricted system access for dial-in users and internal user groups, which prevents access to critical data and system functions?

1949. Are data center personnel trained to recognize hacker activities and patterns?

RETAIL — ROBBERY PREVENTION AND RESPONSE

1950. Does the company provide training for all employees who may face potential robbery situations (employees who have access to or handle cash, checks, and credit cards)?

1951. Is this training given as part of new-hire orientation?

1952. Are employees who handle funds carefully screened?

1953. Are robbery-training situations simulated periodically so that employees may practice their reactions and receive feedback in a safe environment?

1954. Is local law enforcement part of the simulated robbery training?

1955. Have specific procedures been established for the periodic removal of excess funds from all premise areas (registers, cash drawers, manager's office)?

1956. Are all funds transaction areas (registers, payment centers) clearly visible through windows from outside?

1957. Are all panic alarms tested periodically to ensure they are functioning properly?

1958. Is a time-lock safe used with a sign stating that employees do not have access?

1959. Are local law enforcement personnel encouraged to patrol and stop by the premises on a regular basis?

1960. Are signs posted stating that on-hand cash is kept to a minimum?

1961. Are signs posted stating that no bills larger than a $20 will be accepted after designated hours?

1962. Are employees instructed to not openly display large sums of money?

1963. Are emergency phone numbers posted by each telephone?

1964. Are excess cash amounts removed on a random but frequent schedule?

1965. Are employees alert for individuals asking more questions than normal and/or continually hanging around the premises?

1966. Are employees given updates on local area robberies?

1967. Do these updates contain photographs and physical descriptions of the suspects involved?

1968. Is each of the company's locations involved in local business crime prevention groups?

1969. Are employees warned about the dangers of meeting strangers after hours?

1970. Are employees instructed to never leave any keys lying around?

1971. Are all delivery personnel paid by company check rather than cash?

1972. If delivery personnel require cash, is this done in a discreet setting?

1973. During a robbery, are employees told not to try any heroics such as attempting to disarm the perpetrator?

1974. Are employees instructed to do exactly what the robber requests, but nothing more than is specifically demanded?

1975. Are employees instructed to only activate panic alarms if it can be done without the robber noticing?

1976. Are employees instructed to treat all guns or threats of guns as if all are loaded and genuine?

1977. Are employees trained to remember the type and description of the firearm/weapon used?

1978. Are employees instructed to hand over only the amount of money the robber knows about such as a single cash drawer rather than all of them?

1979. If a note from the robber is used, are employees instructed to try to keep it in their possession as evidence?

1980. Are employees instructed to identify anything the robber may have touched?

1981. Are employees instructed to answer only the questions the robber asks and not to ask any of the perpetrator or try to joke around?

1982. As the exception in speaking to the robber, are employees instructed to caution the robber that the employee's supervisor or other person will be arriving at any moment?

1983. Are employees instructed to try to get rid of the robber as quickly as possible?

1984. Are employees instructed to let the robber know exactly what movements are required to gain access to the money such as reaching for a key to open a drawer?

1985. Are employees trained to remember what the robber looked like, especially distinguishing marks (tattoos, scars) or characteristics (accent in voice)?

1986. Are employees trained to determine the direction the robber fled and to get a description of any vehicles involved?

1987. Once the robber has left, are employees instructed to call the police immediately?

1988. Are employees instructed not to hang up the phone until the police have told them to?

1989. While one employee is calling the police, has another employee locked the doors and asked all witnesses to remain inside until the police have interviewed them?

1990. Are employees instructed not to allow anyone else inside the premises unless instructed by the police?

1991. Are employees instructed to fully cooperate with the investigating police agencies?

1992. Are employees instructed to immediately contact the regional or corporate office?

1993. Are employees instructed to prevent anyone from touching anything unless authorized by the police?

1994. If a cash register was robbed, are employees instructed not to process any transactions, which includes closing an open drawer?

1995. Are employees instructed to write down everything they can remember and to have the witnesses do the same?

1996. Are employees instructed to give their statements to the police before they discuss what happened with anyone else?

1997. Are employees instructed to let only a designated company official speak with the media?

1998. Are employees instructed to not reveal the amount of money taken to anyone, especially the media?

1999. Are employees and witnesses allowed to phone their families once the police have finished investigating the scene?

PHYSICAL SECURITY AND OFFICE SECURITY ISSUES

GENERAL SECURITY CONCERNS

2000. Does the company have a security department?

2001. Is the head of the security department an executive level position?

2002. Is the head of security a Certified Protection Professional (CPP) designated by the American Society of Industrial Security?

2003. To whom does the head of security directly report (e.g., the CEO)?

2004. Does the company's management understand and support security's mission?

2005. Does security analyze its programs from a cost to benefit ratio standpoint?

2006. Is this analysis based on recovered assets and reduced or avoided losses?

2007. Does the company offer any incentives to the security budget for cost-saving or loss-control measures?

2008. Is security adequately funded to accomplish its mission?

2009. What are the determining factors for security's budget (e.g., a percentage of sales)?

2010. Is the security department involved with all aspects of operations to help identify vulnerabilities and prevent losses?

2011. Are these vulnerability and loss assessments conducted periodically to adapt to changing conditions?

2012. Has the security department reviewed policies and procedures to determine if they adequately address identified vulnerabilities?

2013. Have the most common risks to personnel been identified (incidence of robbery, hostage situations, etc.)?

2014. Has specific training been designed to address these risks?

2015. Are recognized experts, or their courses, utilized for training personnel to ensure that they act appropriately in high-risk situations such as violent crime or hostage situations?

2016. Does the security department's budget allow for adequate training of both routine and specialized areas?

2017. Does the security department have a formalized training program?

2018. Does the training program address the specific needs at each level and what is required to advance to the next level?

2019. Does each level have clearly defined duties and areas of responsibility?

2020. Does security give awareness training to all of the company's employees?

2021. Has the security department made a complete assessment of all company assets and created short- and long-term plans to protect them?

2022. When a security project is initiated, is a security consultant retained?

2023. Is the security consultant independent of any product or service vendor so as to bias their assessment and solution recommendations?

2024. Have all crimes and security-related incidents, which occurred on company property, been documented and categorized?

2025. Have local law enforcement agencies' crime statistics from the areas surrounding company facilities been gathered and analyzed?

2026. Have crimes and incidents been adequately addressed in security plan strategies?

2027. How many incidents involve physical assets such as inventory and/or office equipment?

2028. How many incidents involve proprietary and competition sensitive information such as sales figures and/or marketing plans?

2029. Is anyone in the security department knowledgeable in preventing computer hackers or, at least, have the knowledge to detect their activities?

2030. Is anyone in the security department knowledgeable about the techniques used for telephone fraud and abuse?

2031. If not, has a reliable expert in computer and telecommunications fraud been identified and utilized?

2032. Does the company have a thorough proprietary information nondisclosure agreement which all employees must sign?

2033. Are there periodic reviews of all security operations to evaluate their effectiveness?

2034. Has the security department developed an informational resource library?

2035. Does the security department regularly receive industry publications such as "Security Management" and "Security" magazines?

2036. Does the security department have a hierarchy?

2037. Are all employees and visitors required to wear identification badges?

2038. Are visitor badges visibly different from employee badges?

2039. Is all badge issuance and tracking controlled by security?

UNIFORMED SECURITY OFFICER STAFF OPERATIONS

2040. Are guards contract employees, proprietary employees, or a combination of the two?

2041. Is the guard force regulated by state law or local ordinance?

2042. Is the guard force commissioned?

2043. What is the total size of the guard force?

2044. Is the guard force unionized? (If so, attach a copy of the current union contract.)

2045. How many guards are on duty per shift?

2046. Is this adequate staffing for the facility needs?

2047. List and describe the organization of the security force.

2048. Who is responsible for formulation of security policy?

2049. How are disciplinary actions for guard infractions enforced?

2050. What responsibilities are covered by the facility's security department?

2051. List the number of shifts utilizing security officers.

2052. List the number of supervisors per shift.

2053. List (attach) responsibilities for guards and guard supervision.

2054. Are all guards familiar with their duties and responsibilities in all major emergency situations? (Interview a cross section of the guard force.)

2055. Are all guards briefed and familiar with the chain of command in normal and all major emergency situations?

2056. Are guards utilized on weekends?

2057. What are the guard reporting procedures for normal and emergency situations?

2058. Who is responsible for following up on guard reports?

2059. Does the guard force have written instructions outlining emergency procedures?

2060. Are guards included in all contingency and drill simulations?

2061. List patrol, Detex key, and foot patrol routes.

2062. Are clock or Detex key stations located appropriately in relation to areas of protection?

2063. List clock or Detex key stations on the facility map.

2064. Are guards encouraged to avoid patterns in the accomplishment of their duties (i.e., timed schedules or key runs)?

2065. How are meal, refreshment, bathroom, and emergency breaks given?

2066. Is the guard force armed?

2067. What weapons are guards issued?

2068. How are the guards trained in the use of these weapons?

2069. List equipment issued to each guard (flash lights, uniform, boots, reflective vests, etc.).

2070. What supplemental equipment is available (tape measures, patrol vehicles, radios, etc.)?

2071. Except for their uniforms, how are the guards identified?

2072. How are disciplinary actions enforced concerning employee infractions against the guards?

2073. Are all employees aware of the authority and responsibilities of the guard force?

2074. How do guards identify items entering and leaving the facility?

2075. Are security posts constructed to allow maximum visibility for security personnel?

2076. Do security officer posts implement increased lighting?

2077. What type of lighting is used in guard post areas?

2078. Is there a system implemented to prevent surveillance of sensitive information through the current communication system?

2079. What is the primary means of guard post communication?

2080. What is the alternate means of guard post communication?

2081. Is there a procedure to identify vendors, subcontractors, and visitors before access is granted to the facility?

2082. Are visitors, maintenance, and vendors required to show identification to the receptionist, guards, or other employees?

2083. Is there an established visitor escort system?

2084. Is a visitors sign-in log maintained?

2085. Is there an established visitor badging system?

2086. What criteria are used for the screening and employment selection for the guard force?

2087. Who is responsible for, and what criteria are used, in the selection of guard supervision/management?

2088. List the turnover and absentee rate for the guard force.

2089. What types of special training are the guards given?

2090. How are guards initially educated to perform their duties?

2091. What type of education is available for guards?

2092. How is continuing education accomplished?

2093. Are guards allowed to attend outside schools and/or education?

2094. Review guard Standard Operating Procedures (SOP), guard manuals, and emergency orders.

2095. Review all contingency plans concerning guard force involvement.

2096. Review procedures for emergency personnel.

PERIMETER BARRIERS AND CONTROLS

2097. Does this facility have a perimeter fence?

2098. Is the perimeter barrier a chain-link fence?

2099. Is the perimeter barrier constructed of solid material? (If so, indicate the location on the facility drawing.)

2100. Is the solid fencing constructed of brick?

2101. Is the solid fencing constructed of metal strips?

2102. Do building perimeters act as a perimeter barrier anywhere in this facility?

2103. Is there a combination of both chain-link fencing and solid material fencing (other than a <u>building</u> perimeter) located as a perimeter barrier in this facility? (If so, indicate the location on the facility map.)

2104. Are natural obstacles part of any portion of the perimeter barrier? (If so, indicate the location on the facility drawing.)

2105. Upon initial inspection, are there any areas of the perimeter fencing or boundaries that present security deficiencies?

2106. Is the fencing fabric constructed of No.6 gauge material?

2107. Is the fencing fabric constructed of No.9 gauge material?

2108. Is the mesh 2 inches apart (indicate if other)?

2109. Excluding the top guard, is the chain-link fence at least 7 feet in height?

2110. Is the chain-link fabric galvanized?

2111. Is the chain-link fabric vinylclad?

2112. Are the corners of fences supported by brace truss rods?

2113. Are wire ties galvanized?

2114. Are fence supports, truss rods, and other fence hardware installed on the interior of the fence?

2115. Are fastening nuts on the interior of the fence spot-welded, peened, or left susceptible to removal?

2116. Are smooth-headed bolts used to attach all fence hardware (lock bolts, gate hardware, stretcher bar bands, etc.)?

2117. Does the fabric reach within 2 inches of hard ground or paving?

2118. Is the fence fabric buried deep enough to compensate for any soil movement?

2119. Is the fence fabric consistently taut?

2120. Are top and bottom fence selvages twisted and barbed?

2121. Are posts equal in height to the fence fabric in all areas?

2122. Is fence fabric tied at 15-inch intervals to the fence posts?

2123. Are the fence post wire ties No.9 gauge?

2124. Are posts positioned no more than 10 feet apart?

2125. Are fence posts constructed of reinforced concrete posts?

2126. Are fence posts constructed of galvanized steel?

2127. Are the supporting posts anchored with concrete?

2128. Is the concrete support "domed" to allow drain-off (shed water)?

2129. Are the supporting posts anchored with metal anchors?

2130. If so, are these strong enough to support the fence properly?

2131. Does the fence have a top rail?

2132. Does the fence have top guards?

2133. Does the top guard increase the height of the fence by at least one foot?

2134. Are the top guards positioned at a 45-degree out-hang?

2135. Are the top guards installed vertically?

2136. Are the top guards positioned at a 45-degree in-hang?

2137. Do the top guards form a "V"?

2138. Do the top guards form an inverted "V"?

2139. Are the top guards collapsible?

2140. Is the top guard 12-gauge galvanized barbed wire with four-point barbs spaced 5 inches or less apart?

2141. If the building or facility walls form part of the perimeter barrier and are less than 2 stories in height, is a top guard attached?

2142. Is the top guard extended an additional 18 inches and sharply angled where perimeter fencing adjoins buildings or facility walls?

2143. Is the fence topped with the standard barbed wire configuration (3 strands, 6 inches apart)?

2144. Is the fence topped with concertina wire?

2145. Is the fence topped with barbed tape?

2146. Is concertina wire positioned at the base of the perimeter fence?

2147. Are barbed tape rolls positioned at the base of the perimeter fence?

2148. Are top guard arms attached to fence posts by spot-welds?

2149. Are top guard arms attached to fence posts by bolting?

2150. List any other top guard information.

2151. Are perimeter openings greater than 96 inches square covered? (If so, indicate location on the facility drawing.)

2152. Is the perimeter fence line alarmed?

2153. Is the fence constructed in a straight line to aid in visual detection of unauthorized individuals?

2154. Are all areas of the fence line illuminated?

2155. Who is responsible for the maintenance of the facility fencing?

2156. How often is this responsibility carried out?

GATE SECURITY AND CONSTRUCTION

2157. Does the perimeter fencing have single-swing gates? (If so, list the number and location on the facility drawing.)

2158. Does the perimeter fencing have sliding track gates? (If so, list the number and location on the facility drawing.)

2159. Does the perimeter fencing have double-swing gates? (If so, list the number and location on the facility drawing.)

2160. Does the perimeter fencing have multifold gates? (If so, list the number and location on the facility drawing.)

2161. Does the perimeter fencing have overhead single-slide gates? (If so, list the number and location on the facility drawing.)

2162. Does the perimeter fencing have double-slide gates? (If so, list the number and location on the facility drawing.)

2163. Does the perimeter fencing have vertical lift gates? (If so, list the number and location on the facility drawing.)

2164. Are there any ingress/egress full-height turnstiles utilized in the perimeter fencing layout? (If so, where are they located?)

2165. Are waist-high turnstiles utilized as access points in the perimeter fencing? (If so, list the number and location on the facility drawing.)

2166. Are gates limited to a functional minimum?

2167. Have all gates been evaluated and determined to be absolutely necessary? If they are not necessary, are they eliminated?

2168. When not being utilized, what procedures are implemented to ensure that all gates remain locked?

2169. Who checks to ensure gates are locked? How is this checked?

2170. Do guards control regularly used gates?

2171. Is gate-area lighting adequate to clearly illuminate all activity?

2172. Are the gates alarmed? (If so, see Intrusion Alarms.)

2173. Are gates the same height as adjacent fencing?

2174. Are gates constructed of the same fabrics as adjacent fencing?

2175. Are gate fabrics secured to anchoring structures in the same manner as adjacent fencing?

2176. Is there no more than 2 inches of space allowed under any gate?

2177. To prevent sagging or excessive flexing, are gates supported by trusses?

2178. What type of gate locking hardware is installed on facility gates?

2179. What type of locks secure gates (combination, pin tumbler, etc.)?

2180. Have locks been attached to the outside gate structure?

2181. Have locks been attached to the inside gate structure?

2182. Are double gate bar-stops secured by a locking device?

2183. Is excessive locked gate travel limited by using an exact chain length?

2184. Are chains secured on gate framing as opposed to fence mesh?

2185. Are top guards placed on all gates?

2186. Are gate top guards the same as top guards on adjacent fences?

2187. Are gate nuts and bolts specifically spot-welded to decrease their vulnerability to attack?

VEHICLE CONTROL AND PERIMETER ENTRY POINT ACCESS

2188. Are all company vehicles equipped with a radio frequency (RF) identification and tracking system?

2189. Are these devices regularly checked and maintained to ensure proper operation?

2190. Can these devices identify each specific vehicle's identification information?

2191. Can these devices interact with the perimeter-point access control system to allow egress?

2192. Can these devices be used to track shipments and integrate with the inventory control system?

2193. If present, are audit trails generated by the vehicle RF sensor identification devices?

2194. Can these devices also be used to track and identify a vehicle if it has been stolen?

2195. Are vehicles inspected when entering the facility?

2196. Are vehicles inspected when exiting the facility?

2197. Is vehicle parking allowed inside the facility's perimeter zone?

2198. To gain access to the facility grounds, are vehicles required to pass through a monitored or supervised entry point?

2199. Do facility personnel and vendors park their vehicles inside the perimeter zone?

2200. Is there a separate area for visitors to park?

2201. Is the visitor parking area located outside or inside the perimeter zone?

2202. Are all parking areas monitored by closed circuit television (CCTV) surveillance during the day and at night?

2203. Are all perimeter gates and entry points to the facility controlled?

2204. Are all perimeter gates and entry points to the facility controlled during normal hours of operation?

2205. Are all perimeter gates and entry points to the facility controlled after normal hours of operation?

2206. Is perimeter point access controlled by an automated access control system?

2207. Are all perimeter gates controlled by an automated access control system?

2208. Are these access control systems activated by swiping an authorization card or by a proximity card?

2209. If a proximity system is used, what is the maximum distance for the card reader?

2210. Does the access control system operate physical barriers such as traffic arms or sliding gates?

2211. Does the access control system record all activity, such as the person, date, and time, which may be converted to a printed report?

2212. Is the access control system integrated with intrusion alarm and CCTV systems for detecting and recording unauthorized entry attempts?

2213. Is the access control system monitored by security personnel?

2214. If an access control system is not used, do the security staff authorize perimeter-point access?

2215. What method does the security staff require to allow entry?

A. Visual recognition of employee

B. Identification badge without employee photograph

C. Identification badge with employee photograph

D. Identification badge with employee photograph verified with authorized roster

2216. Does the security staff require all employees to sign-in before perimeter entry point access is granted?

2217. Does the security staff require all visitors to sign-in, wear a special badge, and be escorted while on the facility?

2218. Are access control systems and authorization rosters immediately updated when there is any change in personnel or their status?

2219. Are perimeter entry points kept to a minimum?

2220. Are the perimeter entry points controlled by cipher locks?

2221. Are the perimeter entry points' cipher lock combinations periodically changed?

2222. If so, who is authorized to change these cipher lock-controlled entry points?

2223. Are the perimeter entry points also controlled by key access?

2224. If so, who is authorized to have keys?

2225. Are cipher lock and key access perimeter entry points monitored by CCTV, security personnel, and/or intrusion alarms?

2226. Are all personnel exits from the facility monitored and recorded?

2227. Do facility employees challenge anyone who is not properly escorted or not wearing proper identification while they are still in the perimeter zone?

GROUNDS SECURITY AND PROTECTIVE CLEAR ZONES

2228. Is there a “clear zone” on the exterior side of the facility’s perimeter fence?

2229. Is there a “clear zone” on the interior side of this facility’s perimeter fence?

2230. Has the exterior “clear zone” been designed to be a minimum width of 50 feet?

2231. Has the interior “clear zone” been designed to be a minimum width of 20 feet?

2232. Is there a clear vehicle path for guard patrols located on the outside of the perimeter fence or barriers?

2233. Are visual obstructions such as tall grass removed from the view of the “clear zone?”

2234. Are “clear zone” areas adequately illuminated?

2235. Could the perimeter fence be scaled with any materials or items in the close proximity of the area such as construction material, buildings, tall trees, etc.?

2236. Are all crates, merchandise, and stock kept away from the perimeter fence line and building areas?

2237. To prevent the obstruction of the “clear zone” view, has all vegetation overgrowth been removed from the perimeter fence line?

2238. To enhance security surveillance, has the amount and the height of shrubbery near windows, doors, gates, fence lines, and access roads been limited?

2239. Are all subterranean passageways, overpasses, or raised walkways near the perimeter fence secured?

2240. Have special procedures been implemented to protect blind alleys which are adjacent to buildings?

2241. Have prominent warning signs been posted at 100-feet intervals on the property's perimeter?

2242. From a moderate distance can warning signs be easily read?

LIGHTING ASSESSMENT

2243. Are all facility roads adequately illuminated?

2244. Is the area just beyond the facility's "clear zone" properly illuminated to make surveillance easier?

2245. Is the area just inside the facility's "clear zone" area properly illuminated?

2246. Does lighting system use overlapping cones of light to illuminate the perimeter areas?

2247. Is lighting evenly dispersed to avoid high-contrast areas?

2248. Is vegetation kept away from lighting to prevent masking?

2249. Are portable lights available if needed?

2250. Are light levels periodically tested with a sensitive light meter?

2251. Are all perimeter light fixtures located on the interior of the perimeter fence?

2252. Is the lighting adequate to illuminate critical facility areas?

2253. Is there an even distribution of light throughout the facility?

2254. What physical systems protect entry points to the perimeter lighting system?

2255. Are regularly scheduled tests performed on protective lighting?

2256. Who performs these periodic tests?

2257. Are facility light fixtures protected against tampering and vandalism?

2258. Can protective lighting be reached and therefore compromised by unscrewing bulbs?

2259. Are all lighting system timers, switch box(es), and photo-sensitive cells secured?

2260. Is normal lighting supplemented with sensor-activated additional lighting?

2261. What are the procedures for cleaning and replacing lamps?

2262. Who routinely performs these procedures?

2263. If a critical area luminary burns out, who replaces it?

2264. How is the protective lighting system initiated?

2265. What are the operational hours of protective lighting systems?

2266. Is daytime lighting adequate for security purposes?

2267. During periods of poor visibility, does the facility's protective lighting operate continuously?

2268. Is there a glare factor hampering the security force's view anywhere within the facility?

2269. Does the facility's lighting produce a spectrum of light that will be effective for CCTV surveillance?

2270. Does the current security lighting system provide the desired levels of illumination at a reasonable cost?

2271. Does the protective lighting system have an emergency back-up power source which is easily initiated?

2272. List the emergency power back-up source for protective lighting currently being used.

2273. Where is the emergency power back-up system located?

2274. Who is responsible for maintenance of the emergency power back-up system?

2275. List the Foot-candles or lux for major critical facility areas.

CLOSED CIRCUIT TELEVISION (CCTV)

2276. Does your facility use CCTV surveillance systems to augment other alarm and security measures?

2277. Is the current system adequate for existing needs?

2278. Is the employee parking lot monitored by CCTV cameras?

2279. Is the visitor parking area monitored by CCTV cameras?

2280. Are these areas illuminated properly for CCTV purposes?

2281. Are indoor and outdoor cameras strategically positioned to maximize effectiveness?

2282. How many cameras are attached to pan/tilt drives?

   A. Where are these located?

   B. Who services them?

   C. Indicate each camera location on the facility map.

2283. Is there more than one pan/tilt/zoom (P/T/Z) control station?

2284. If using multiple P/T/Z control stations, is there a priority system or is it first come, first served?

2285. Are CCTV cameras mounted in a manner that creates an overlapping effect?

2286. Do these CCTV cameras have sufficient ambient light for proper operation? (If not, what type of artificial light is used, is it adequate, etc.?)

2287. How many CCTV cameras utilize the following?

A. Wide-angle lenses

B. Infrared filters

C. Low-light capabilities

2288. Are critical areas monitored by CCTV cameras?

2289. Are exterior cameras adequately protected from the environment (heat, cold, dust, etc.)?

2290. Do camera assemblies have adequate protection from tampering?

2291. Are all cable runs protected from tampering, including tamper switches on junction boxes?

2292. Is the power cable of adequate size to supply enough current to all devices (camera, heater, blower, wiper, controller, infrared illuminator, etc.)?

2293. Is the power cable installed in a separate raceway from the video signal cable (coax) to avoid electrical interference?

2294. Is power to all camera assemblies on the same electrical circuit to avoid phase problems?

2295. Are circuit breakers marked and locked to avoid tampering?

2296. Are the cameras phase-adjustable?

2297. Are all cameras at the same ground potential?

2298. Does the system use ground loop isolators if using coaxial cable?

2299. If coaxial cable is used, are there any sizes other than RG-59U (RG-11, RG-6, etc.)?

2300. Do coaxial cables use a copper core and braiding for shielding?

2301. Are there any other video transmission technologies used (fiber optics, microwave, RF, etc.)?

2302. Is the CCTV system monochrome, color, or both?

2303. Is there more than one monitoring location?

2304. Are video signals looped through the monitors or split using a distribution amplifier?

2305. Who visually monitors the CCTV system monitors (guards, reception personnel, etc.)?

2306. Are the monitors mounted at an angle to avoid glare on the screen?

2307. Is the monitoring area(s) designed with adequate ventilation to dissipate heat from the CCTV equipment?

2308. Can the CCTV monitoring/control equipment be easily accessed for repair/maintenance?

2309. Does the CCTV monitoring system have video tape recording ability?

2310. Does taping exclusively occur during alarm events or is it continually recorded by time lapse recorder?

2311. Is there a separate playback monitor for viewing tape recordings?

2312. Does the CCTV system also interface with other systems (intrusion, access control, etc.)?

2313. Is the interface hardware (relays, etc.) or software (integration)?

2314. Describe the CCTV system maintenance program. (Who is responsible for maintenance, how are damaged cameras repaired, etc.?)

ACCESS CONTROL

2315. Does the company utilize electronic access control?

2316. List the specific areas where these systems are located.

2317. What type(s) of access reading systems are utilized (digital, data chip, card, biometrics, etc.)?

2318. If digital access control is used, are the pass codes common or unique to each employee?

2319. If unique codes are used, are the numbers selectable?

2320. If digital access control is used, are the codes kept in a secure location?

2321. If digital access control is used, does the reader include a duress or panic feature?

2322. Does the system have the capability to combine a digital access control and card reader?

2323. Does the system have the capability of changing from digital access control to card reader with both using time zones?

2324. If card reader, what type of card is issued (barium ferrite, optical scan, Wiegand, magnetic stripe, or proximity)?

2325. Is an employee ID badge integrated with the access control card?

2326. Is the employee's picture on the card?

2327. Is the picture a laminated photograph or digital picture?

2328. Are access control cards issued to visitors on a temporary basis?

2329. Is blank ID badge stock secured and inventoried?

2330. If segregated access is required in your facility, are badges color-coded to allow compartmentalized access?

2331. Are all badges serially numbered and recorded as to whom they were issued to?

2332. Is the responsibility for the control and issuance of badges under the auspices of the security department? (If not, who is responsible?)

2333. How is lost badge replacement accomplished?

2334. If badges are lost, what is the investigative process that follows?

2335. Does the access control system use any biometric technologies?

2336. Which technology is utilized: hand geometry, fingerprints, retinal scan, voice, signature, or keystroke?

2337. Is the system a PC-based system?

2338. What operating system does the PC use (DOS, Windows, UNIX, etc.)?

2339. Can the system make database changes while still monitoring alarms (multitasking)?

2340. Does the system have any remote workstations (multiuser)?

2341. What type of network does the system use to connect to remote workstations?

2342. Does the system software include a self-diagnosis to detect hardware or software malfunctions?

2343. Does the system have the capability for remote diagnostics via modem?

2344. Does the system have access to a help screen via a single "hot-button"?

2345. Does the system have a context-sensitive help screen?

2346. Does the system have a backup for all software (operating system and database)?

2347. What type of backup technology is used (e.g., floppy disks, secondary hard drive, tape, floptical, etc.)?

2348. Does the system have the capability to assign unique fields to the database?

2349. What type of database does the system use (hierarchical, network, relational, SQL)?

2350. Does the system have the capability for a graphical map display?

2351. Is the system capable of importing existing map display graphics?

2352. What format of graphics can be imported (bit map, vector, both)?

2353. Can the system provide anti-passback?

2354. Can the system provide global anti-passback?

2355. Can the system provide linking (an alarm on one processor activating a relay on another)?

2356. Can the system provide global linking?

2357. Can the system provide elevator control?

2358. How many cabs and floors can be controlled?

2359. Can the system supervise a guard tour?

2360. How many tours and stations can be supervised?

2361. Can the system provide time and attendance?

2362. Can the system assign a priority to each alarm signal and display them in order of priority?

2363. How many time zones can the system assign?

2364. How many door groups can the system assign?

2365. Can the system use a mouse rather than a keyboard for executing most commands?

2366. Does the system allow users to menu-select options from a pop-up list?

2367. How many readers can be used with the system?

2368. How many cards or codes can be used with the system?

2369. How many alarm-inputs can be used with the system?

2370. How many alarm-inputs can be used with each processor?

2371. Are alarm-inputs supervised (end-of-line resistor) to detect opens, shorts, ground, etc.?

2372. How many relay outputs can be used with the system?

2373. How many relay outputs can be used with each processor?

2374. Does the system use distributive processing?

2375. How many transactions can be stored at each processor?

2376. How many hours of battery back-up does each processor have?

2377. How many operators can be logged on the system?

2378. Can the operators be identified by name instead of number?

2379. How many operator privilege levels does the system provide?

2380. Does the system provide a menu to choose which events will be printed and at which printer?

2381. How many alarm-inputs can be used with each processor?

2382. Does the access control system's doors use electric strike and/or magnetic locks?

2383. Does the control unit, readers, and locks receive power from a single source?

2384. Are the circuit breakers for the access control system labeled?

2385. Does the system rely on a battery backup during a power failure?

2386. Do access controlled door contacts communicate an open or closed status to the system monitoring function?

2387. Are the contacts separate or are they part of the electric lock?

2388. Do the controller doors also have a latch-bolt monitor?

2389. Are "hold open" alarms installed on these doors?

2390. Are alarm shunt relays installed on all readers?

2391. Is a tamper or forced-entry relay installed on all access-controlled doors?

2392. Has an exit button or sensor been installed on the interior of access-controlled doors to allow personnel to exit?

2393. What type of exit device is used (push button, touch bar, motion detector, beam, mat, etc.)?

2394. Is there a system-generated audit trail which tracks the access point, time, and user number, and whether access was granted or denied?

2395. Is a printed report generated documenting the system's activities?

2396. Is this report transferred to a computer disk for storage?

2397. Is all user information and access levels continually kept updated?

2398. Who is responsible for performing this function, and how frequently is it completed?

2399. Can a user's access be immediately removed or changed at any time if circumstances warrant this action?

2400. During emergency situations, are there special procedures for the access control system to maintain its integrity?

2401. Do installed systems meet all current building and fire codes? (If not, which ones do not?)

2402. Does the system limit access to an area equivalent to its appropriate level of threat?

2403. Is the type and amount of personnel use considered in the design capabilities of the access control system?

2404. Is exposure to environmental factors taken into consideration when the system is being designed?

2405. Are system components scheduled for periodic service?

2406. What is the access control system's manufacturer recommendation for various maintenance services?

2407. Can company personnel perform these maintenance functions, or is a vendor's services required?

2408. If a vendor service is required is the vendor escorted or signed in?

2409. Does the access control system have any alarm features?

2410. If alarmed, does the system sound locally and at a remote area or site?

2411. Do security personnel monitor the system?

2412. Do security personnel respond to any system alarms?

2413. Is there a specifically required response time to any alarms?

LOCKING SYSTEMS AND PROCEDURES

2414. Are there established procedures for controlling locking systems?

2415. Who controls and oversees the locking system program?

2416. How many keys have been issued?

2417. Of the total number of keys issued, how many are master keys?

2418. Of the total number of keys issued, how many are grand-master keys?

2419. Are keys issued on the basis of a person's job function and level of responsibility?

2420. Are inventories performed to determine the number of keys issued and who conducts them?

2421. Are keys held to an absolute minimum?

2422. When was the last inventory conducted?

2423. Were all keys accounted for?

2424. If not, how many were missing?

2425. Who are lost keys reported to if other than the key control officer?

2426. Are key holders prohibited from duplicating keys for any reason?

A. Where is the company's key duplication facility?

B. Is "Do Not Duplicate" stamped on all keys issued to discourage duplication?

2427. Who is responsible for issuing replacement keys, if other than the key control officer?

2428. Are duplicate keys stored in a separate key security control box?

2429. Are all facility keys accounted for?

2430. How are keys accounted for?

2431. Who accounts for facility keys, if other than the key control officer?

2432. Is a fireproof and lockable key-control security box used to store keys not in use and key blanks?

2433. Is the distribution and use of the keys to the key-control security box kept to an absolute minimum?

2434. List all the individuals with a key who has access to the key-control security box.

2435. Are keys allowed off-site?

2436. Is there a key issuance audit book which records all activity?

2437. Is a secure area used to store the key issuance audit book?

2438. Who is responsible for reviewing the accuracy of the key issuance audit book?

2439. Is it strictly prohibited to leave keys unattended for any period of time?

2440. Who has access to master keys?

2441. Are master keys devoid of any master key markings?

2442. Are master keys issued on a sign-out basis?

2443. Are there specific procedures for assuring the return of facility keys from suspended, resigning, or terminated employees?

2444. Who performs the locksmith duties?

2445. Does the facility employ a certified locksmith?

2446. If not, has a certified locksmith been contracted to provide services?

2447. Are all facility locks inspected periodically to ensure functionality and to detect tampering?

2448. When was the last time facility locks were changed?

2449. Who is responsible for facility lock inspection to test their functionality?

2450. Are all locks currently functioning properly?

2451. Has access to keys and combinations been strictly limited to only personnel who need them in the performance of their job duties?

2452. What criteria are used to determine key and combination access requirements for individual employees?

2453. To maintain and protect combination integrity, what procedures and safeguards are used?

2454. Is there a "sign-off" procedure or record log requiring signatures for all individuals with combination access?

2455. Are combinations to "sensitive information" containers changed at least every 6 months?

2456. Are combinations changed on discharge, suspension, or reassignment of any individual having knowledge of the combination?

2457. Regardless of known security violations and personnel transfers, are locks changed at one-year intervals at a minimum?

2458. Is a combination immediately changed if an incident or situation has compromised it?

2459. Are cipher lock viewing shields used to deny unauthorized observation of combinations?

2460. When they are not being used, do all doors remain securely locked?

2461. When not in use, are all padlocks secured to a hasp to prevent unauthorized removal?

2462. Are all bolts installed in such a manner that it prevents them from being easily cut?

2463. If bolts are not protected, can an offset or coverguard plate be added to secure the bolt area?

2464. Are all screws and holding locks firmly secured in place?

2465. Are door locks and frames equipped with metal jimmy guards to help prevent forcible spreading attack?

INTRUSION ALARMS

2466. Is the facility equipped with intrusion alarms?

2467. Who is the manufacturer of the alarm systems?

2468. Can the system pinpoint the status of each sensor?

2469. Can pinpoint annunciation be performed at each arming station or only at the main control panel?

2470. Can the system disarm or bypass sensors in groups by their respective areas?

2471. Does the company own its alarm system?

    A. If not, is a leased alarm system agreement in effect?

    B. Which alarm leasing company is used?

    C. Does the lease include a maintenance contract?

2472. Is an off-site central station used to monitor the facility's alarm system?

2473. Is the alarm system monitored by direct police connection?

2474. Is the monitoring of the alarm system proprietary (in-house)?

2475. Does the alarm system sound locally?

2476. What type of audible signal is used (bell, siren, voice, etc.)?

2477. Does the system also use visual signaling (strobe lights, etc.)?

2478. Does the system produce an audible or visual alarm throughout the entire facility or just in the area of intrusion?

2479. Does the alarm system use current technology? When was it installed?

2480. Has Underwriters Laboratories, Inc. (UL) approval been given to all alarm-system components?

2481. Has the alarm system been designed to be overt or covert to the public?

2482. Are replacement components, such as fuses, batteries, sensors, etc., for the alarm system kept on-site?

2483. Is alarm system deactivation only possible from inside a secured area?

2484. Are exposed alarm system components protected by tamperproof fixtures and/or tampering alarms?

2485. Does the facility contain exterior intrusion detection devices?

2486. If so, what type of sensors are used? (e.g., gate contacts, beams, motion detection, buried cable, etc.)

2487. Are all exterior sensors specifically designed for exterior use?

2488. Are all exterior sensors protected from the environment as much as possible?

2489. Are exterior sensors scheduled for seasonal sensitivity adjustments?

2490. Is the alarm system equipped with an alternative power source in case of a main power failure?

2491. What is the alarm system's alternative power source?

2492. Are tests performed to ensure that the alarm system is operating properly?

2493. How frequently are these tests performed?

2494. Who is responsible for performing these tests (company personnel and/or vendors)?

2495. What was the date of the last test conducted on the facility's alarm system?

2496. Is maintenance on the alarm system performed by company personnel or vendors?

2497. If alarm system maintenance is done by outside vendors, have emergency repair provisions been established?

2498. Is a log of all alarm system false alarms and failures accurately kept?

2499. Are supervised transmission lines used for alarm system communications?

2500. Are perimeter building walls protected with vibration detectors or capacitance alarms?

2501. How is the interior space above false ceilings protected (volumetric alarms, etc.)?

2502. If these false ceilings lead to critical areas, are they protected by intrusion alarms and CCTV cameras?

2503. Are ventilation and HVAC ducts protected with barriers, inspection ports, break-wire alarms, acoustic baffles, duct traps, capacitance, or other alarm sensors?

2504. Is the guard response time to alarm activation adequate?

2505. Is the position of the alarm enunciator panel adequate for quick recognition of alarm conditions?

2506. Is the alarm system equipped with a "remote test" function that lets you analyze each zone as to its functionality?

2507. If an ultrasonic or volumetric sensor is used, has the oscillator circuit been modified to notify the enunciator panel (to create an alarm condition) if a sensor fails?

OFFICE AREA SECURITY

2508. Does the facility have a periodic security survey and review conducted by a Certified Protection Professional (CPP) or outside security consultant?

2509. Is the office protected by a monitored alarm system?

2510. Are pass codes to the alarm system strictly controlled?

2511. Is a nightly inspection performed by the last individual to leave the office to ensure that all doors and windows are locked and that no one is hiding inside?

2512. Is there an established procedure to ensure the proper functioning of the alarm system nightly (in each office area)?

2513. Who activates the office alarm system?

2514. If your facility does not incorporate an alarm system, is there a guard force present?

2515. Is the office protected by a CCTV system?

2516. Is the CCTV system operating continuously?

2517. Is the CCTV system being recorded on videotape?

2518. If a multiple camera system is being used, is the recording system equipped with a multiplexer to capture all images on one videotape?

2519. Are some cameras in sensitive areas utilizing video motion detectors which can trigger an intrusion alarm and record activity on videotape at the same time?

2520. Are all triggered alarms immediately responded to, thoroughly reviewed, and the tapes kept as evidence?

2521. Is there adequate tape stock rotation to allow time to carefully review all tapes before they are reused to ensure that no alarms or other suspicious events are erased?

2522. Are all cameras of solid state circuit (CCD) design?

2523. Is lighting in all areas adequate for the cameras to display and record clear and easily visible pictures?

2524. Are employees situated to be able to easily view entrances, stairs, and elevators?

2525. Are all doors secured after the normal hours of operation to protect employees when they are working late or working extra hours?

2526. Are emergency numbers for local emergency medical services, police, and fire departments easily accessible to employees?

2527. Are PCs, typewriters, printers, and other high value movable items bolted down?

2528. Are items of value removed from desks or unsecured in some manner?

2529. When the office is left unattended during the day, are lockable file cabinets, desks, and other containers secured?

2530. Is information removed from all desktops during off-hours and placed into fireproof locked cabinets or safes?

2531. Who is responsible for checking desk areas during off-hour times?

2532. Are attaché and briefcases carrying valuable information stored in a secure manner when not in use?

2533. Are filing cabinet keys immediately returned to the key-control security box after the cabinets are opened at the beginning of the business day?

2534. Are filing cabinet locks supplemented by an exterior locking bar to give additional security?

2535. Before important or company-sensitive papers are discarded, are they shredded?

2536. Are office machines such as computers, photocopiers, typewriters, and calculators secured by the use of a sturdy lock or locking device?

2537. Is office equipment marked or tagged, inventoried, and periodically audited?

2538. Are all equipment serial numbers recorded and placed in a locked file cabinet or safe?

2539. Is the equipment serial number list, a description of each item, and photograph or video inventory of these items placed in the safe and duplicate copies of all materials given to the company's insurance provider?

2540. To prevent unauthorized toll calls, can the telephone system be locked by an electronic or manual mechanism?

2541. If not, are lockers with company-provided locks furnished to employees?

2542. Is there a special secured cabinet for all office-related keys?

2543. When given out, are keys tracked by an issuance log which requires that the person requesting the key sign for it?

2544. Who is responsible for key control, tracking, and issuance?

2545. Are keys only issued to those employees who have a legitimate purpose for having them?

2546. Are key issuance log and other information current and complete?

2547. Are employees prohibited from making duplicate copies of any company keys?

2548. Does the company use a patented lock and key system to reduce the ability to copy the keys?

2549. Is it prohibited to leave any key in a locking file cabinet or any lock?

2550. Are procedures established to prevent an unauthorized person from obtaining a replacement key by reporting a missing key (file cabinets, office machines, desks, etc.)?

2551. Are daily deposits made to prevent large amounts of cash and checks from being left at the facility overnight?

2552. Are funds in the safe kept to an absolute minimum?

2553. Where is the safe kept?

2554. Is the safe UL listed for both fire and burglary resistance?

2555. Is the safe attached to a floor or wall structure?

2556. Is the safe combination known by only a minimal number of employees, and is it changed periodically?

2557. Is it mandatory that the safe lock be completely reset every time it is used rather than being left in a "day-lock" mode?

2558. Is the safe equipped with a keypad combination lock which can record all activity by individual pass code numbers?

2559. Is the safe combination always changed immediately when an employee who had access is terminated?

2560. Is the safe equipped with a time-lock feature?

2561. Is the safe always checked to ensure that it is locked as part of the daily closing procedure?

2562. Are the number of doors limited to those that are absolutely necessary?

2563. When areas are not in use, are the access doors kept locked?

2564. Are doors securing exterior and "sensitive area" office areas secured with double cylinder deadbolt and a key-locked doorknob?

2565. Are all doors constructed of sturdy metal with metal frame?

2566. If wood doors are utilized, are they of a solid construction?

2567. Do doors have a minimum fire rating?

2568. Are all door hinges located on the inside of the door frame?

2569. Have door hinge pins been secured by bradding or spot-welding?

2570. Are all windows capable of being locked?

2571. Are window locks supplemented with additional devices such as a secondary locking system and antisliding bars?

2572. Are ventilation and other openings properly secured from the interior and exterior?

2573. Are roof hatches and skylights locked and protected by the alarm system?

2574. Are emergency telephone numbers listed throughout the office?

2575. Is the office area patrolled by security officers?

2576. Does this patrol cover the exterior and interior office areas?

2577. If there is an interior patrol, are the patrol officers using a station and time recording system such as a Detex clock?

2578. Is this patrol record carefully reviewed and audited?

2579. Is the office building perimeter lighting sufficient?

2580. Is the office building parking lot lighting sufficient?

2581. Are exterior lights controlled by photosensing devices or by a timer?

2582. Are office building entrances and hallways sufficiently lit?

2583. Is after-hours interior lighting controlled by closing procedures or by a timing system?

2584. Is lighting routinely inspected for proper operation?

2585. Are all burnt-out bulbs immediately replaced?

2586. Are all lighting devices protected by vandal-resistant fixtures?

2587. Are all employees issued an identification badge?

2588. Does the badge contain a photograph and the name of the employee?

2589. Is the badge devoid of any obvious markings which denote the employee's position or department?

2590. Is the badge also used to activate access control systems?

2591. Is the badge designed and constructed so that it would be extremely difficult to duplicate?

2592. Are employees cautioned not to wear their badges in public?

2593. If a badge is lost, is its access authorization immediately eliminated?

2594. When a replacement badge is issued, is a new badge number used rather than the old one?

2595. Are all visitors required to sign-in at the receptionist's desk and receive a badge?

2596. Does the visitor log track the name of visitors, the date, time-in, time-out, and the person who was visited?

2597. Are visitor badges noticeably different from regular employee badges?

2598. Are visitor badges always collected when the visitor is leaving and has signed out in the visitor log?

2599. Is the visitor escorted at all times by the designated employee?

2600. Are all employees required to immediately assist anyone they observe walking around the facility without a badge?

2601. Is the issuance and control of badges handled by the security department?

2602. Does human resources notify security prior to the termination of an employee?

INVENTORY CONTROL

2603. Does top management support the security department and the inventory control program?

2604. Is an Electronic Asset Protection (EAP) system used to control inventory?

2605. Are EAP tags attached to all items of value such as computers, printers, telephones, merchandise, etc.?

2606. Are EAP tags attached to company tools, machinery, and equipment to help reduce attempted thefts?

2607. Is a CCTV system integrated with the EAP system to control items in critical areas?

2608. Is the EAP system interfaced with the access control system to help prevent a potential theft from after-hour surreptitious exiting?

2609. Is the EAP system integrated with the intrusion alarm system?

2610. Are all of these systems combined in a configuration to allow them to be monitored simultaneously from a single console?

2611. Has a recent inventory of business equipment been completed, listing serial numbers and descriptions?

2612. When conducting a department's inventory, are employees outside that department utilized?

2613. How often is a complete inventory conducted?

2614. Are continual inventories conducted to identify missing items as well as patterns of theft?

2615. How are missing or lost items reported?

2616. Where are "found" items stored? (Does the facility have an established lost and found area?)

2617. Are all items critical to your company's operation secured (computers, etc.)?

2618. Is there a log maintained recording the serial numbers of all valuable company equipment?

2619. Are there adequate controls over usable scrap and waste, company tools, and equipment used by employees?

2620. Are valuables or highly desirable items maintained in secure areas?

2621. Are competitive proposals and bids, reviewed by management?

2622. Is there a policy against using only single-source suppliers?

2623. Is there a written policy warning employees against accepting gifts from vendors?

2624. Are copy machines locked at the end of each workday?

2625. When locked, are copy machines' meters recorded to ensure material is not copied?

2626. Is there a written company policy allowing security or management to inspect desks, file cabinets, and other possible storage containers as well as employee-owned vehicles parked on company property?

2627. Is there a package or personnel item inspection program (e.g., lunch bags, briefcases and purses), in effect throughout the facility?

2628. If there is a package inspection in effect, is there a package release system? (If so, is there a list of authorized release signatures at each egress point?)

2629. How often are these lists updated? (What is the updating procedure to ensure that all needed signatures are listed?)

2630. Are "inbound" packages inspected for authorization?

2631. Are all boxes, crates, and containers inspected before leaving the facility?

2632. Are vehicles exiting the facility inspected?

2633. Are railroad cars inspected before exiting the facility?

2634. Are bills of lading inspected and reconciled against actual cargo before it exits from the facility?

2635. If transportation seals are used, are they checked against the manifest?

2636. Are parking or staging areas for trucks adequately controlled to prohibit theft of company items?

2637. How is the removal of trash and trash receptacles controlled?

2638. Is there a written policy on company theft describing disciplinary action for theft or vandalism?

2639. Does the employee hiring process normally require a background investigation to determine if there is a history of theft?

2640. Are employees prohibited from parking near receiving areas?

2641. Are desirable items in the shipping and receiving process secured?

2642. When transferring items in the shipping and receiving areas, is a count taken each time there is a transfer? (Who supervises this activity?)

2643. Are all trucks sealed after a last count is taken before exiting the shipping and receiving area?

TRANSPORTATION SECURITY

2644. Are shipping and receiving doors kept closed and locked unless in use?

2645. Are doors to the shipping and receiving areas alarmed?

2646. Are shipping and receiving areas physically separated as much as functionally possible?

2647. Is the shipping and receiving facility located on private property, not a regular public roadway?

2648. Are keys to all gates, doors, trucks, containers, and special holding areas tightly controlled and accounted for at all times?

2649. Are all trucks and containers clearly marked with the company's name and identification numbers on all sides?

2650. Are cargo containers sealed for large shipments?

2651. Do only a limited number of people have access to the seals for cargo shipments?

2652. Are seals routinely inventoried?

2653. Are all seals inspected to verify they are intact when shipments are arriving or leaving at the dock?

2654. Are seal numbers and shipment numbers tracked together?

2655. Are outgoing shipments clearly labeled and contain the appropriate order number?

2656. Are labels designed to stay attached to their shipment?

2657. Are labels generated by the sales department, not the shipping department?

2658. Are shipping documents and products shipped together?

2659. After a shipment is received and verified, are shipping documents checked-off as received, and is the date/time stamped?

2660. When a product is being received, is its actual physical count compared against the shipping documentation and the company's purchase order?

2661. Are items secured if they must be stored for a period of time?

2662. If a shipment contains a number of loose boxes, have proper measures been taken to secure them until they are shipped out?

2663. Are any freight-damaged products received immediately placed in a secured area to await claim processing and preserving evidence?

2664. Does the facility have photographic equipment to document damaged products and/or the associated delivery personnel?

2665. Are all receiving and shipping employees fidelity insurance bonded?

2666. Are personnel responsible for outbound and inbound shipments required to sign all relevant documentation?

2667. Are several employees always used to ship and receive products as a check and balance system?

2668. Is there a manager responsible for overall warehousing operations?

2669. Do supervisors make random and unannounced audits of shipping and receiving procedures?

2670. Are surprise physical inventories of storage areas also conducted?

2671. Do supervisors routinely work the floor areas of the warehouse operation?

2672. Are unauthorized individuals prohibited in company vehicles at any time?

2673. Are company drivers warned that any deviations from the assigned route are strictly forbidden?

2674. Are the identities of consignment drivers verified before they are allowed into the warehouse facility?

Appendix A — Glossary of Terms

/bin - A major directory of the UNIX operating system. Short for command binaries directory. The name of the directory where most executable public programs and its utilities of the UNIX operating system reside.

/etc - A major directory of the UNIX operating system, which consists of system configuration files and executables. This directory contains many administrative and configuration files.

/etc/groups - On most UNIX systems, the text file which associates user accounts into groups with different privileges.

/etc/hosts - On most UNIX systems, the text file which associates IP addresses with host names, including alias names for the host. This file provides access to host systems by alias name without having to enter an IP address. This file is located in the /etc directory and is called hosts.

/etc/password - The directory name of where the password file is stored in most flavors of the UNIX Operating System.

10BaseT - IEEE's specification for running Ethernet over unshielded twisted-pair wiring.

2600 - A hacker organization who publishes 2600 magazine.

8lgm - 8 Little Green Men hacker group that puts out security tips.

A

Abuse of Privilege - When a user performs an action that they should not have according to organizational policy or law.

Access Port - A logical or physical identifier that a computer uses to distinguish terminal input/output data streams.

ACL - Access Control List; a list normally used for controlling access to computer programs and services. Identifies the users who may access an object and the type of access to the object that a user is permitted.

ActiveX - Microsoft scripting language used for their Internet browser.

AFCERT - Air Force Computer Emergency Response Team; located at the Air Force Information Warfare Center, Kelly AFB, Texas.

AIS - Automated Information System; any equipment or interconnected system or subsystems of equipment that is used in the automatic acquisition, storage, manipulation, control, display, transmission, or reception of data and includes software, firmware, and hardware.

Alert - A formatted message describing a circumstance relevant to network security. Alerts are often derived from critical audit events.

Anonymous FTP - An anonymous file transfer protocol that will accept any user name and an anonymous or no password. This is a common way for hackers to gain access to a computer. This allows a user to log-on a remote machine as anonymous, without having a user account on that system.

ANSI - American National Standards Institute.

Anti-Virus Program - Software program designed to protect an AIS from a virus attack.

Application Gateway - One form of a firewall. A firewall looks for valid application-level data before allowing a connection. An ftp server for example appears as an ftp server to the client and as an ftp client to the server.

Archie - This was once a powerful service; greatly expands the usefulness of anonymous ftp up until the World Wide Web was established. This service helps locate the file, program, or other information that you are looking for.

ARP (Address Resolution Protocol) - A protocol within the Transmission Control Protocol/Internet Protocol (TCP/IP) suite that maps IP addresses to Ethernet addresses. TCP/IP requires ARP for use with Ethernet.

ARPANET (Advanced Research Projects Agency Network) - A network originally sponsored by DARPA to link universities and government research centers. The TCP/IP protocols were pioneered on ARPANET.

ARQ - A control code that calls for the retransmission of a block of data.

ASCII (American Standard Code for Information Interchange) - The data alphabet used in the IBM PC to determine the composition of the 7-bit string of 0s and 1s that represents each character (alphabetic, numeric, or special).

ASIM (Automated Security Incident Measurement) - Automated security tool that monitors network traffic and collects information on targeted unit networks by detecting unauthorized network activity.

Assessment - Surveys, Inspections; An analysis of the vulnerabilities of AIS. Information acquisition and review process designed to help a customer determine how best to use resources to protect information in systems in order to be successful in the mission.

Assurance - A measure of confidence that the security features and architecture of an AIS accurately mediate and enforce the security policy.

Asynchronous - A method of transmission in which the time intervals between characters do not have to be equal. Start and stop bits are added to coordinate the transfer of characters.

ATM - Asynchronous Transfer Mode.

Attack - An attempt to bypass security controls on a computer. An active attack alters data. A passive attack releases data. Whether an attack will succeed depends on the vulnerability of the computer system and the effectiveness of existing countermeasures.

Audit - The independent examination of records to access their accuracy and completeness.

Audit trail - In computer security systems, a chronological record of when users log-in, how long they are engaged in various activities, what they were doing, and whether any actual or attempted security violations occurred. An automated or manual set of chronological records of system activities that may enable the reconstruction and examination of a sequence of events and/or changes in an event.

AUSCERT - Australian Computer Emergency Response Team.

Authenticate - In networking, to establish the validity of a user or an object (i.e., communications server).

Authentication - To positively verify the identity of a user, device, or other entity in a computer system, often as a prerequisite to allowing access to resources in a system.

Authorization - The process of determining what types of activities are permitted. Usually, authorization is in the context of authentication. Once you have authenticated a user, the user may be authorized different types of accesses or activities.

Automated Security Monitoring - All security features needed to provide an acceptable level of protection from hardware, software, and classified sensitive unclassified or critical data, material, or processes in the system.

Availability - Computer hardware and software system working efficiently, and the system is able to recover quickly and completely if a disaster occurs.

awk - A rudimentary (UNIX) programming language used mainly with text and database files through editing, sorting, and searching.

B

Back Door - A hole in the security of a computer system deliberately left in place by designers or maintainers. Synonymous with trap door; A hidden software or hardware mechanism used to circumvent security controls. A breach created intentionally for collecting, altering, or destroying data.

Backup - Copy of files and programs made to facilitate recovery if necessary.

Banner - On interactive software, a first screen containing a logo, author credits, and/or a copyright notice. DoD computers are required to have a warning banner.

BASH - Bourne Again Shell. This is a UNIX shell.

Bastion hosts - A system that has been hardened to resist attack and which is installed on a network in such a way that it is expected to potentially come under attack. Often are components of firewalls.

Baud - Bits per second. A measure of transmission speed.

BBS - Bulletin Board System; a menu-driven program which provides communications, e-mail messaging, uploading, and downloading of software. Typically used by amateurs for fun at home. Considered the low-rent district for the hacker culture.

BIOS - Basic Input/Output System.

BIP - Base Information Protection office.

bit - The unit of information. A computational quantity that can take on one of two values, such as 0 and 1 or true and false.

BLOB - Binary Large Object; used to describe any random large block of bits, usually a picture or sound file. Can be stored in a database but normally not interpretable by a database program. Can be used as a mild hacker threat (mailbomb) when mailed. Can also be used to hide malicious logic code.

Bomb - A general synonym for crash; normally of software or operating system failures.

Boot - To load and initialize the operating system on a machine.

Bounce - An electronic mail message that is undeliverable and returns an error to the sender.

BPS - Bits per second.

Breach - The successful defeat of security controls; could result in a penetration of the system. A violation of controls of a particular information system such that information assets or system components are unduly exposed.

Bridge - A functional unit that interconnects two local area networks that use the same control procedure.

Broadcast - To send a message to all stations or an entire class of stations connected to the network.

Brouter - A device that combines the functions of a bridge and a router. Brouters can route one or more protocols, such as TCP/IP and XNS, and bridge all other traffic.

Browser - A program designed to help users access, view, and navigate on the Internet (World Wide Web). Browsers are sometimes also called Web clients since they get information from a server. Examples are Netscape, Mosaic, Microsoft Explorer, and Apache.

Brute Force - A primitive programming style (ignorance); one in which the programmer relies on the computer's processing power instead of using his or her own intelligence to simplify the problem, often ignoring problems of scale and applying naive methods suited to small problems directly to large ones.

Buffer - A temporary storage space. Data may be stored in a buffer as it is received before or after transmission. A buffer may be used to compensate for the differences between the speed of transmission and the speed of processing.

Buffer Overflow - This happens when more data are put into a buffer or holding area then the buffer can handle. This is caused by a mismatch in processing rates between the producing and consuming processes.

Bulletproof - Technical speak for describing an algorithm or implementation considered extremely robust; loss-resistant; capable of correctly recovering from any imaginable exception condition — a rare and valued quality. Same as armor-plated.

Bug - An unwanted or unintended property of a program or piece of hardware that causes it to malfunction.

Bus - Transmission path or channel.

Bus topology - A broadcast arrangement in which all network stations receive the same message through the cable at the same time.

Byte - A unit of memory or data equal to the amount used to represent one character. For most computers this is usually 8 bits.

C

C - A programming language used predominantly by professional programmers to write applications software. The name of a programming language designed by Dennis Ritchie during the early 1970s and immediately used to reimplement UNIX; so called because many features derived from an earlier compiler named `B' in commemoration of its parent, BCPL.

C++ - An object-oriented programming language. An improved version of the C programming language.

Cache - An amount of RAM set aside to hold data that is expected to be accessed again. The second access, which finds the data in RAM, is very fast.

Cache Memory - A special buffer storage, smaller and faster than main storage, that is used to hold a copy of instructions and data in main storage that are likely to be needed next by the processor.

Call Back Security - Procedure for identifying a remote AIS terminal, whereby the host system disconnects the caller and then dials the authorized telephone number of the remote terminal to re-establish the connection.

cat - Short for catalog. A UNIX command to list out (non-stop scrolling) the contents of a text file on the screen.

cd - A DOS and UNIX command to change into a different directory of a file system.

CDS - CSAP Database System. An AFCERT relational database system containing information such as detailed computer hardware and software specifications, vulnerabilities and countermeasures, malicious logic, and system connectivity descriptions.

CERT - Computer Emergency Response Team; supports others in enhancing the security of their computing systems; develops standardized set of responses to security problems; provides a central point of contact for information about security incidents; assists in collecting and disseminating information on issues related to computer security, including information on configuration, management, and bug fixes for systems.

Certificate - A relatively new option for client authentication; forces use of password protection at the beginning of a session and can be setup to prompt for password while logged in.

Certificate Authority (Workstation) - Establishes and manages user keys, privileges and certificates. Records the security information about an IS user. A workstation where certificates are generated, stored, protected, transferred, loaded, used, and destroyed.

CGI (Common Gateway Interface) - Method that Web servers use to allow interaction between servers and programs.

CGI Scripts - Allows for the creation of dynamic and interactive web pages. They also tend to be the most vulnerable part of a web server (besides the underlying host security).

Channel - A path between sender and receiver that carries one stream of information.

Chat Group - A virtual meeting place where you can converse with other users from all parts of the globe. The chat groups are "live."

Check_Password - A hacking program used for cracking VMS passwords. See Guess_Password.

Chernobyl Packet (Kamikaze Packet) - A network packet that induces a broadcast storm and network meltdown. Typically an IP Ethernet datagram that passes through a gateway with both source and destination Ether and IP address set as the respective broadcast addresses for the subnetworks being gated between.

chmod (change mode) - A UNIX command to change the read, write, and execute privileges for files.

chown (change owner) - A UNIX command to change the ownership of a file from one user to another.

chgrp (change group) - A UNIX command to change the group accessing a file from one group to another.

chrootuid - A software tool which runs network services with a lower authorization than root. Restricts file system access of network services.

CIAC (Computer Incident Advisory Capability) - An organization of the Department of Energy which provides computer security services.

Circuit Level Gateway - One form of a firewall. Validates TCP and UDP sessions before opening a connection. Creates a handshake, and once that occurs it passes everything through until session is ended.

COAST (Computer Operations, Audit, and Security Tools) - Organization at Purdue University which collects computer security tools.

Common Carrier - A transmission company that serves the general public (telephone company).

Communications Security - Procedures designed to ensure that telecommunications messages maintain their integrity and are not accessible by unauthorized individuals.

Compartmented Mode Workstation (CMW) - A secure workstation providing a high-resolution monitor, a window manager, and a detailed set of security functions. CMWs have enough built-in security to operate as trusted computers.

compress - A UNIX command to compress a file, creating a filename with a .Z extension.

Compress/compression - When used without a qualifier, it generally refers to crunching of a file using a particular algorithm for minimizing the amount of space needed for the files to be stored.

Compromise - An intrusion into a computer system where root level access (superuser privileges) is gained and files have been altered or changed.

Computer Abuse - The willful or negligent unauthorized activity that affects the availability, confidentiality, or integrity of computer resources. Computer abuse includes fraud, embezzlement, theft, malicious damage, unauthorized use, denial of service, and misappropriation.

Computer Emergency Response Team (CERT) - Supports others in enhancing the security of their computing systems; develops standardized set of responses to security problems; provides a central point of contact for information about security incidents; assists in collecting and disseminating information on issues related to computer security, including information on configuration, management, and bug fixes for systems.

Computer Fraud - Computer-related crimes involving deliberate misrepresentation or alteration of data to obtain something of value.

Computer Geek - One who eats computer bugs for a living. One who fulfills all the stereotypes about hackers.

Computer Security - Technological and managerial procedures applied to computer systems to ensure the availability, integrity, and confidentiality of information managed by the computer system.

Computer Security Incident - Any intrusion or attempted intrusion into a computer system. Incidents can include probes of multiple computer systems.

Computer Security Intrusion - Any event of unauthorized access or penetration to a computer system.

concentrator - See wiring hub.

Connectionless Data Transfer (Mode Transmission) - In packet data transmission, a mode of operation in which each packet is encoded with a header containing a destination address sufficient to permit the independent delivery of the packet without the aid of additional instructions. A connectionless packet is frequently called a datagram. A connectionless service is inherently unreliable in the sense that the service provider usually cannot provide assurance against the loss, error, insertion, or misdelivery.

Control Zone - The space expressed in feet of radius, surrounding equipment processing classified information which is under sufficient physical and technical control to preclude a successful hostile intercept attack.

Cookie - A handle, transaction ID, or other token of agreement between cooperating programs.

COPS (Computer Oracle and Password System) - A computer network monitoring system for UNIX machines. Software tool for checking security on shell scripts and C programs. Checks for security weaknesses and provides warnings.

Copy Protection - A class of methods for preventing incompetent pirates from stealing software and legitimate customers from using it.

Core - The main storage of a computer system. A core dump captures data and files in memory. Main storage or RAM. Dates from the days of ferrite-core memory; now archaic.

Core Dump - A copy of the contents of core memory produced when a process is aborted by certain kinds of internal error.

COTS Software - Software acquired by government contract through a commercial vendor. This software is a standard product, not developed by a vendor for a particular government project.

Countermeasures - Action, device, procedure, technique, or other measure that reduces the vulnerability of an automated information system. Countermeasures that are aimed at specific threats and vulnerabilities involve more active techniques as well as activities traditionally perceived as security.

cp - A UNIX command to copy files.

CPS - Character per second.

CPU (Central Processing Unit) - The functional "brain" of the computer; the element that does the actual adding and subtracting of 0s and 1s that is essential to computing. Normally the main integrated circuit located on a motherboard in a computer.

Crack - A popular hacking tool used to crack passwords. System administrators also use Crack to assess weak passwords by novice users in order to better secure his/her system.

Crack Root - To defeat the security system of a UNIX machine and gain root system privileges thereby.

Cracker - One who breaks security on a system.

Cracking - The act of breaking into a computer system.

Crash - A sudden, usually drastic failure of a computer system.

crypt - A UNIX command that converts a text file and stores it in a new encrypted format.

Cryptographic Checksum - A one-way function applied to a file to produce a unique "fingerprint" of the file for later reference. Checksum systems are a primary means of detecting file system tampering on UNIX.

CSAP (Computer Security Assistance Program) - Program to implement information protection operations capabilities using a combination of administrative controls, reporting procedures, specially developed automated security tools, incident response, intelligence threat data, and special survey and analysis capabilities.

CSET (Computer Security Engineering Team) - The CSET focus on specific systems and networks. The team overtly surveys policy, procedures, training, awareness, connectivity and physical security of a specific system. CSETs document problems with policy and directives, operating procedures, configuration management, training and awareness, system connectivity, physical security, and unauthorized software.

Cyberspace - Information-space of a computer system, or system of networks. The metaphoric location of the mind of a person in hack or cracking mode.

D

Daemon - A program that is executed automatically on the completion of a specific operation on the knowledge base. A program that is not invoked explicitly, but lies dormant waiting for some condition to occur before it reacts.

Dark-side Hacker - A criminal or malicious hacker.

DARPA - Defense Advanced Research Projects Agency. Now known as ARPA.

Data driven attack - A form of attack that is encoded in innocuous appearing data, which is executed by users or other software, to implement an attack. In the case of firewalls, a data driven attack is a concern since it may penetrate the firewall in data form and launch an attack against a system behind the firewall.

Data Encryption Standard (DES) - An encryption standard developed by IBM and then tested and adopted by the National Bureau of Standards. Published in 1977, the DES standard has proven itself over nearly 20 years of use in both government and private sectors. Cryptographic algorithm designed for the protection of unclassified data and published by the National Institute of Standards and Technology.

Data Stream - A sequence of digitally encoded signals used to represent information for transmission.

Deadlock - A situation wherein two or more processes are unable to proceed because each is waiting for another to do something. See Deadly Embrace.

Deadly Embrace - Same as Deadlock, though usually used only when exactly two processes are involved.

Demon Dialer - A program which repeatedly calls the same telephone number. This is benign and legitimate for access to a BBS or malign when used as a prank for denial of service attack.

Denial of Service - Action(s) which prevent any part of an AIS from functioning in accordance with its intended purpose.

Denial Time - The average length of time that an affected asset is denied to the organization.

Derf - The act of exploiting a terminal, which someone else has absentmindedly left logged on

DES - Data Encryption Standard. A national cryptographic standard based on a 56-bit algorithm. This algorithm can be implemented in electronic hardware devices and used for the cryptographic protection of computer data.

Designated Approving Authority (DAA) - Official with the authority to formally assume responsibility for operating a system at an acceptable level of risk.

Dial-up line - A communications circuit established by dialing a destination over a commercial telephone system.

Dial-Up Security - The service whereby a computer terminal can use the telephone to initiate and effect communication with a computer.

dig - A service quite similar to nslookup for querying and checking names of an IP address.

diskless client - A system with no local disk drive that relies on an NFS server for the operating system, swap space, file storage, and other basic services.

DNS (Domain Name Service) - A distributed networked-based naming service on the Internet that associates IP addresses with the name of a user site.

DNS spoofing - Assuming the DNS name of another system by either corrupting the name service cache of a victim system or by compromising a domain name server for a valid domain.

Domain - A directory structure for electronic mail addressing and network address naming. Within the United States, top-level domains include .com, .edu, .gov, .mil, .net, and .org. Subdomains designate the organization and the individual system.

Down - Not working.

Download - To transfer data or code from one system to another, usually over a digital communication link.

Dragon - A program similar to a "daemon," except that it is not invoked at all, but is instead used by the system to perform various secondary tasks. A typical example would be an accounting program, which keeps track of who is logged in.

Dual homed gateway - A dual homed gateway is a system that has two or more network interfaces, each of which is connected to a different network. In firewall configuration, a dual homed gateway usually acts to block or filter some or all of the traffic trying to pass between the networks.

Dump - An undigested and large amount of information routed to an output device. Usually it is a backup of computer files and data.

Dumpster Diving - The practice of raiding the dumpsters behind buildings where producers and/or consumers of high-tech equipment are located with the expectation of finding discarded but still-valuable equipment or information.

Duplex - In communication circuits, the ability to transmit and receive at the same time; also referred to as full duplex. See half duplex.

E

ed - A standard UNIX text editor. If the filename given doesn't exist, it creates it.

Edit - A UNIX line-oriented text editor that runs a simplified version of ex for novice users.

Elevator seeking - A method of optimizing the movement of the heads on the hard disk in a file server.

Elm - A mail reader for the UNIX operating system.

E-mail/electronic mail - An electronic means for communicating information (primarily text) by a method of sending, storing, processing, and retrieving the information. Messages are held in storage until called for by the addressee.

emacs (EMACS) - Short for Editing MACroS. A UNIX text editor complete with customizable programming. It includes program with compilation subprocesses and send and receive mail features. There are many variants created mainly by hackers for most major operating systems. The most widely used UNIX version is called GNU EMACS. Approximately 80% of hackers online time is spent using EMACS. It is considered to be the ultra editor for hackers.

Emulation - Simulation of a system , function, or program.

Encryption - The process of scrambling files or programs, changing one character string to another through an algorithm (such as the DES algorithm).

env - A UNIX command to display the current environment or, if environment variables are specified, set them to a new value and display the contents.

Ethernet - A network cable and access protocol scheme originally developed by Xerox; now marketed mainly by Cisco, Bay, 3Com, and DEC.

Ethernet sniffing - This is listening with software to the raw Ethernet device for packets that interest the user. When the software sees a packet that fits certain criteria, it logs it to a file. The most common criteria for an interesting packet is one that contains words like login or password.

ex - A UNIX line-oriented text editor; a superset of ed and the root of vi.

F

FAQ (Frequently Asked Question) - A collection of common questions.

Fault Tolerance - A method of ensuring continued operation through redundancy and diversity.

Fiber Optics - A data-transmission method that uses light pulses sent over glass cables.

File Server - A type of server that holds files in private and shared directories for LAN users.

File System - A hierarchical arrangement of directories and files.

File Transfer Protocol (FTP) - Used for transmitting files between systems on a network. See FTP.

Filter - A program that processes an input data stream into an output data stream in some well-defined way.

Finger - A program that displays information about a particular user or all users logged on the system or a remote system.

Firewall - A system or combination of systems that enforces a boundary between two or more networks. Gateway that limits access between networks in accordance with local security policy. The typical firewall is an inexpensive microbased UNIX box kept clean of critical data, with a bunch of modems and public network ports on it but just one carefully watched connection back to the rest of the cluster.

Firewall machine - Computer on a network used to isolate, filter, and protect local systems from external connectivity by controlling the amount and kinds of traffic that will pass between the two. A dedicated gateway machine with special security precautions on it used to service outside network connections and dial-in lines. The idea is to protect a cluster of more loosely administered machines hidden behind it from crackers.

Fishbowl - To contain, isolate, and monitor an unauthorized user within a system to gain information about the user.

Fix - A patch or solution to a reported computer problem.

Flame - An online message intended to insult or provoke other users.

Flood - To dump large amounts of text onto an Internet Relay Chat. This is especially rude when the text is uninteresting and the other users are trying to have a serious conversation.

Flooding - Sending lots of text to the screen at once.

Fork Bomb (Logic Bomb) - Code that can be written in one line of code on any UNIX system; used to recursively spawn copies of itself, "explodes" and eventually eating all the process table entries and effectively locks up the system.

FORTEZZA - Converter for the PCMCIA card containing cryptographic algorithm, keys, and related functions developed for Mosaic and Netscape.

Freeware - Free software distributed by e-mail or bulletin boards.

Fried - Non-working due to hardware failure; burnt out.

Front-End Security Filter - Security filter logically separated from the remainder of an IS to protect system integrity. Synonymous with firewall.

FTP (file transfer protocol) - A protocol by which a connection can be made between two computers and information can be exchanged; downloaded or uploaded.

G

Gateway - An interface between two networks. A system that handles e-mail traffic between different communication networks.

get - an FTP command to "get" or download a file from a remote machine.

GID number - The group identification number used by the system to control access to accounts owned by other users.

GIF - Graphical Image Format; an image file.

Glitch - A sudden interruption in electric service, sanity, continuity, or program function.

GNU - A UNIX-workalike development effort of the Free Software Foundation.

Good Times - A hoax virus.

Gopher - A type of Internet service, now almost obsolete due to the World Wide Web. A menuing interface to a tree or graph of links to documents, executable programs, or other menus across the net. See Veronica, Archie, and Jughead.

grep – a UNIX command to globally search for the regular expression or particular string or pattern in a file or set of files and print the lines containing matches to it.

Guard - A processor that provides a filter between two systems operating at different security levels or between a user terminal and a database to filter out data that the user is not authorized to access.

Guess_Password - A hacking program used for cracking VMS passwords. See Check_Password.

GUI - Graphical User Interface.

Gun - To forcibly terminate a program or job. See Kill.

Guru - An expert. Implies not only wizard skill but also a history of being a knowledge resource for others.

H

Hack - Short for hacker. A quick job that produces what is needed, but not well. To pull a prank on.

Hack Attack - Synonymous with hacking run, though more strongly implies an all-nighter.

Hack Mode - What one is in when hacking, of course.

Hacker - A person who enjoys exploring the details of computers and how to stretch their capabilities. A malicious or inquisitive meddler who tries to discover information by poking around. A person who enjoys learning the details of programming systems and how to stretch their capabilities, as opposed to most users who prefer to learn only the minimum necessary.

Hacking - Unauthorized use or attempts to circumvent or bypass the security mechanisms of an information system or network.

Hacking Run - A hack session extended long outside normal working times, especially one longer than 12 hours.

Half Duplex - Alternating transmissions; each station can either transmit or receive, not both simultaneously.

Handshaking - Hardware or software activity designed to start or keep two machines or programs in synchronization as they do protocol.

Hang/Hung - To wait for an event that will never occur. Equivalent to wedged.

Hardwired - In software, it is synonymous for hardcoded. By extension, anything that is not modifiable, especially in the sense of customizable to one's particular needs or tastes.

Heartbeat - The signal emitted by an Ethernet transceiver at the end of every packet to show that the collision-detection circuit is still connected. A periodic synchronization signal used by software or hardware, such as a bus clock or a periodic interrupt.

Hog - Favored term to describe programs or hardware that seem to eat far more than their share of a system's resources, especially those which noticeably degrade interactive response.

Home Box - A hacker's personal machine, especially one he or she owns.

Home directory - The part of the file system that is allocated to an individual user for private files.

Home Page - A World Wide Web (WWW) repository for information.

Hook - A software or hardware feature included to simplify later additions or changes by a user.

Hose - To make a system or software nonfunctional or greatly degrade in performance.

Host - Computer providing processing power for attached terminals and peripheral devices. The controlling computer of a network.

HTML (Hyper Text Markup Language) - The standard programming language of the World Wide Web. Computer language and the tags used by web browsers.

HTTP (Hyper Text Transport Protocol) - Protocol that is used to request and receive documents from servers on the World Wide Web.

Hypertext - Data that contains links to other data.

Hz - Cycles per second.

I

ICMP (Internet Control Message Protocol) - The TCP/IP process that provides the set of functions used for network-layer management and control.

IDEA - International Data Encryption Algorithm.

Identification and Authentication - Process of establishing the validity of a claimed identity. Generally accomplished by the use of unique machine-readable user names.

Identity Token - A smart card, a metal key, or some other physical token carried by a systems user that allows user ID validation.

IEEE - Institute of Electrical and Electronics Engineers.

Incident - Event that has actual or potentially adverse effects on AIS. Any intrusion or attempted intrusion into a computer system. Incidents can include probes of multiple computer systems.

inetd - This is the UNIX Internet services daemon. It is the server process for the Internet standard service. It is normally started up during boot time.

Infection - The act or result of affecting injuriously; an infective agent or material contaminated with an infective agent, usually malicious logic in the form of a worm, virus, Trojan horse, etc.

Information Systems Security Officer (ISSO) - The person responsible to the DAA for ensuring that security is provided for and implemented throughout the lifecycle of an AIS.

Information Security - The result of any system of policies and procedures for identifying, controlling, and protecting from unauthorized disclosure; information whose protection is authorized by executive order or statute.

Information Systems Security (INFOSEC) - Protection of information systems against unauthorized access to or modification of information, whether in storage, processing, or transit and against the denial of service to authorized users.

Information Warfare - Actions taken to achieve information superiority by affecting adversary information, information-based processes, and information systems, while defending our own information, information-based processes, and information systems. Any action to deny, exploit, corrupt, or destroy the enemy's information and its functions, protect ourselves against those actions; and exploiting our own military information functions.

Integrated Services Digital Network - Standardized operating parameters and interfaces for a network that will allow mixed digital transmission services simultaneously.

Internet - A collection of worldwide computer networks.

Internet Address - Any network address reachable through the Internet, normally identifiable with an absolute network address in the form username@sitename.subdomain.domain.

Internet Outdial - An Internet outdial is a modem connected to the Internet and can be used to dial out. Normal outdials will only call local numbers. A GOD (Global Outdial) is capable of calling long distance. Outdials are an inexpensive method of calling long distance BBSs.

Internet Protocol (IP) - A standard describing software that keeps track of the Internet address for different nodes, routes outgoing messages, and recognizes incoming messages. A DoD standard protocol designed for use in interconnected systems of packet-switched computer communication networks.

Internet Service Provider (ISP) - ISPs sell Internet access to the mass market. While the big nationwide commercial services with Internet access (like America Online, CompuServe, Genie, Netcom, etc.) are technically ISPs, the term is usually reserved for local or regional small providers (often run by hackers turned entrepreneurs) who resell Internet access cheaply without themselves being information providers or selling advertising.

Internet Worm - Independent program that replicates from machine to machine across network connections often clogging networks and computer systems as it spreads.

InterNIC - The Internet Registry for host registration services and contains Internet name services.

Intrusion - An unauthorized access or penetration of a computer system.

Intrusion Detection - Pertaining to techniques which attempt to detect intrusion into a computer or network by observation of security logs or audit data. Detection of break-ins or attempts either manually or via software expert systems that operate on logs or other information available on the network.

Investigation - Review and analysis of system security features to determine the security provided by the operating system.

I/O - input/output.

IP address - A unique Internet protocol number that identifies each system in a network.

IP slicing/hijacking - An attack whereby an active, established, session is intercepted and coopted by the attacker. IP splicing attacks may occur after an authentication has been made, permitting the attacker to assume the role of an already authorized user. Primary protections against IP splicing rely on encryption at the session or network layer.

IP spoofing - An attack whereby a system attempts to illicitly impersonate another system by using an IP network address.

IRC (Internet Relay Chat) - A worldwide "party line" network that allows one to converse with others in real time. IRC is structured as a network of Internet servers, each of which accepts connections from client programs, one per user.

ISDN (Integrated Services Digital Networking) - A network that shares high-speed data communications along with voice communications on the same wires. It splits basic service into shared channels, each of which can be used for different calls.

ISP - See Internet Service Provider.

J

Java - A computer language developed by Sun for web browsers.

Joe Password - A type of logging into a computer system where the password is the same as the user name.

JPEG - An image file.

Jughead - A software tool similar to Veronica, except that it searches only a confined area of gopherspace. See Veronica, Archie, and Gopher.

Jupiter - To kill an Internet Relay Chat robot or user and then take its place by adopting its nickname or handle so that it cannot reconnect. Named after a particular IRC user who did this to NickServ, the robot in charge of preventing people from inadvertently using a nick claimed by another user.

K

Kerberos - A "trusted third party" authentication system developed at MIT; allows clients and servers to authenticate and communicate securely over an untrusted network. It uses symmetric-key encryption technology (DES) to avoid passwords being sent over the network in the clear.

Kernel - The hardware, firmware, and software elements of a trusted computing base that implement the reference monitor concept. It must mediate all accesses, be protected from modification, and be verifiable as correct.

Key - A symbol or sequence of symbols which control encryption and decryption.

Key Card - Paper card containing a pattern of punched holes which establishes the key for a specific cryptonet at specific time.

keylogin - The UNIX command to decrypt and store a secret key with keyserv. The command prompts for a password, and uses it to decrypt the user's secret key. These encryption keys are used for accessing secure network services such as secure NFS and NIS+.

keyserv - The UNIX command for storing private encryption keys. It is a daemon that is used for storing the private encryption keys of each user logged into the system.

Keystroke Monitoring - A specialized form of audit trail software or a specially designed device that records every key struck by a user and every character of the response that the host computer returns to the user.

Kill - To forcibly terminate a program or job.

L

LAN (Local Area Network) - A computer communications system limited to no more than a few miles and using high-speed connections (2 to 100 megabits per second). A short-haul data communications system that connects ADP devices in a building or group of buildings with a few square kilometers, including workstations, front-end processors, controllers, switches, and gateways.

Leapfrog Attack - Use of userID and password information obtained illicitly from one host to compromise another host. The act of TELNETing through one or more hosts in order to confuse a trace (a standard cracker procedure).

Letterbomb - A piece of e-mail containing live data intended to do malicious things to the recipient's machine or terminal. Under UNIX, a letterbomb can also try to get part of its contents interpreted as a shell command to the mailer. The results of this could range from silly to tragic.

Link - A word or picture you select on a Web page, usually by clicking your mouse on it, leading to a new page.

Linux - The free UNIX workalike created by Linus Torvalds and friends starting about 1990. This may be the most remarkable hacker project in history -- an entire clone of UNIX for 386, 486, and Pentium micros, distributed for free with sources over the net.

ln - A UNIX command to create pseudonyms (links) for files, allowing them to be accessed by different names.

Logic Bomb - A resident computer program which, when executed, checks for particular conditions or particular states of the system which, when satisfied, triggers the perpetration of an unauthorized act.

login - A UNIX command to log into a system.

lp, lpr - A UNIX command to send files to the printer.

ls - A UNIX command to list out directory contents; with a multitude of options and variations.

M

mail - A UNIX command to read mail or send mail to other users.

Mail Storm - What often happens when a machine with an Internet connection and active users reconnects after extended downtime — a flood of incoming mail that brings the machine to its knees.

Mailbomb – A technique that forwards a massive amount of e-mail to a single system or person with the intent to crash the recipient's system. Mailbombing is widely regarded as a serious offense as it is an extremely effective denial of service attack.

mailx - A UNIX command used to send and receive mail to other users. Some SysAdmins configure systems with a variety of mail services, such as elm, mailtool, dtmail, etc. See Sendmail.

Malicious Code - Hardware, software, or firmware that is intentionally included in a system for an unauthorized purpose; e.g., a Trojan horse.

Malicious Logic - Hardware, software, or firmware that is intentionally included in a system for an unauthorized purpose; e.g., a Trojan horse. It is intentionally included in an IS for an unauthorized purpose.

man - A UNIX command (short for manual) to get help information.

Mimicking - Spoofing.

MISSI (Multilevel Information Systems Security Initiative) Component - Any platform that is FORTEZZA or CC equipped.

Mockingbird - A computer program or process which mimics the legitimate behavior of a normal system feature (or other apparently useful function) but performs malicious activities once invoked by the user.

Modem - A device that modulates and demodulates signals. Modems are primarily used for converting digital signals into quasi-analog signals for transmission over analog communication channels and for reconverting them back into digital signals. Many additional functions may be added to a modem to provide for customer service and control features. Modems may be internal or external to the hardware of a computer.

more - A UNIX command to display the contents of a text file on the screen, one screen at a time. See cat.

MUD (Multiple User Dimension) - A computer program that provides a virtual reality. One can telnet to a MUD server, take on a role, and explore and interact with other users who are playing their roles.

Multi-Level - Concept of processing information with different classifications and categories that simultaneously permits access by users with different security clearances, but prevents users from obtaining access to information for which they lack authorization.

Multilevel Security - Concept of processing information with different classifications and categories that simultaneously permits access by users with different security clearances but prevents users from obtaining access to information for which they lack authorization.

Multitasking - A mode of operation that provides for concurrent performance or interleaved execution of two or more tasks.

Multiuser Operation - A system with the capability to have multiple users accessing and processing simultaneously such as is found with UNIX.

mv - A UNIX command to move a file from one location to another; can be used to rename a file as well.

N

Nak Attack - A penetration technique which capitalizes on a potential weakness in an operating system that does not handle asynchronous interrupts properly and thus leaves the system in an unprotected state during such interrupts.

Nastygram - A protocol packet or item of e-mail (like letterbomb) that takes advantage of misfeatures or security holes on the target system.

Network - A continuing connection between two or more computers that facilitates sharing files and resources.

Network Access Points (NAP) - These are connections between ISPs which can cause bottlenecks in throughput.

Network Address - As used by hackers, means an address on the network that is almost always an Internet address.

Network Level Firewall - A firewall in which traffic is examined at the network protocol packet level.

NFS (Network File System) - One of many distributed-file system protocols that allow a computer on a network to use the files and peripherals of another networked computer as if they were local. This protocol was developed by Sun Microsystems and adopted by other vendors.

NIS (Network Information System) - This is the current name for what was once known as yp (Yellow Pages). The purpose for NIS is to allow many machines on a network to share configuration information, including password data. NIS is not designed to promote system security.

nispasswd - UNIX command to change the password information stored in Network Information Service Plus Version 3 (NIS+).

node - A connection or switching point on the network.

nslookup - A service for querying and checking names of an IP address.

Nuke - To intentionally delete the entire contents of a given directory or storage volume.

O

On-line - Connected to a network or a host computer system.

OLS (On-line Survey) - Remote computer survey of known technical vulnerabilities through security tools.

One-Time Passwords - Passwords that are changed after each use and are useful when the password is not adequately protected from compromise during log-in. A private character string that is used only once to authenticate an identity. After each use, a new character string is generated.

Operational ASIM site - ASIM has been installed, is capturing and analyzing data, and is sending the data to the AFCERT.

Orange Book - The U.S. Government's standards document "Trusted Computer System Evaluation Criteria, DOD standard 5200.28-STD, December, 1985" which characterizes secure computing architectures and defines levels A1 (most secure) through D (least). Stock UNIXs are roughly C1, and can be upgraded to about C2 without excessive pain.

P

Packet - A block of data sent over the network transmitting the identities of the sending and receiving stations, error-control information, and a message.

Packet Filter - Inspects each packet for user-defined content, such as an IP address but does not track the state of sessions. This is the least secure type of firewall.

Packet Filtering - A feature incorporated into routers and bridges to limit the flow of information based on predetermined communications such as source, destination, or type of service being provided by the network. Packet filters let the administrator limit protocol specific traffic to one network segment, isolate e-mail domains, and perform many other traffic control functions.

Packet Sniffer - A device or program that monitors the data traveling between computers on a network.

Packet Switched Network - A network of devices that communicate between each other by transmitting packets addressed to particular destinations. A transmission technique that maximizes the use of digital transmission facilities by transmitting packets of digital data from many customers simultaneously on a single communications channel.

Packet Switching - The process of routing and transferring data by means of addressed packets so that a channel is occupied during the transmission of the packet only and on completion of the transmission the channel is made available for the transfer of other traffic.

Partition - A discrete portion of a disk, configured using the format program in UNIX, and Fdisk in DOS.

Passive Attack - Attack which does not result in an unauthorized state change, such as an attack that only monitors and/or records data.

Passive Threat - The threat of unauthorized disclosure of information without changing the state of the system. A type of threat that involves the interception, not the alteration, of information.

Passphrase - A sequence of characters, longer than the acceptable length of a password, that is transformed by a password system into a virtual password of acceptable length.

passwd - A UNIX command to change a user's password.

Password - A protected word or string of characters that identifies or authenticates a user for access to a specific resource such as a system, data set, file, record, and so forth.

Password Shadowing - Password shadowing is a security system where the encrypted password field of /etc/passwd file is replaced with a special token, and the encrypted password is stored in a separate file which is not readable by normal system users.

Patch - A temporary addition to a piece of code, usually as a quick remedy to an existing bug or misfeature.

Patterns - Stereotyped operations, procedures, or processes that reveal protected information.

PBX (Private Branch Exchange) - A telephone system serving a specific location. Many PBX systems can carry computer data without the use of modems.

Penetration - The successful unauthorized access to an automated system.

Penetration Signature - The description of a situation or set of conditions in which a penetration could occur or of system events, which in conjunction can indicate the occurrence of a penetration in progress.

Peon - A person with no special root or wheel privileges on a computer system.

Perimeter-based security - The technique of securing a network by controlling access to all entry and exit points of the network.

Perl - Practical Extraction and Report Language. An interpreted computer language used mostly with UNIX systems.

Permissions - A description of the type of authorized interactions a subject can have with an object.

perpetrator - The entity from the external environment that is taken to be the cause of a risk. An entity in the external environment that performs an attack.

PGP (Pretty Good Privacy) - An encryption program.

phage - A program that modifies other programs or databases in unauthorized ways; especially one that propagates a virus or Trojan Horse.

phf - Phone book file demonstration program that hackers use to gain access to a computer system and potentially read and capture password files.

phf hack - A well-known and vulnerable CGI script which does not filter out special characters (such as a new line) input by a user.

Phracker - Individual who combines phone phreaking with computer hacking.

Phreak(er) - Individual fascinated by the telephone system. Commonly, an individual who uses his knowledge of the telephone system to make calls at the expense of another.

Phreaking - The art and science of cracking the phone network.

Piggy Back - The gaining of unauthorized access to a system via another user's legitimate connection.

Pilot Error - A user's misconfiguration or misuse of a piece of software, producing apparently bug-like results.

Ping (Packet Internet Groper) - An exercise program associated with TCP/IP and used to test the Internet communications channel between stations. The slang term for a small network message sent by a computer to check for the presence and alertness of another computer.

Ping of Death - The use of Ping with an address number higher than 65,507. This will cause a SYN flood and cause a denial of service. RFC-791 says IP packets can be up to 65,535, with the IP header of 20 bytes, and ICMP header of 8 octets (65,535-20-8 = 65,507). Sending a bigger packet greater than 65,507 octets causes the originating system to fragment the packet.

Pizza Box - The thin box housing the electronics in desktop workstations; so named because of its size and shape and the dimpled pattern that looks like air holes.

Pkzip - A program to compress a files.

Polling - A method of controlling the transmission sequence of communicating devices on a shared circuit by sending an inquiry to each device asking whether it wishes to transmit.

Power hit - A spike or drop-out in the electricity supplying your machine.

Priority Interrupt - Describes any stimulus compelling enough to yank one right out of hack mode.

Privileged Access - Explicitly authorized access of a specific user, process, or computer to a computer resource.

Privileges - Pertaining to a program or user and characterized the type of operation that can be performed. A right granted to a user, a program, or a process.

Probe - Any effort to gather information about a machine or its users on-line for the apparent purpose of gaining unauthorized access to the system at a later date.

Process - A program in operation. In UNIX, the ps command will display all processes currently running.

Profile - Patterns of a user's activity which can detect changes in normal routines.

Program - A plan or routing for solving a problem on a computer. A sequence of instructions used by a computer to do a particular job or solve a given problem.

Promiscuous Mode - When an Ethernet interface reads all traffic (sniffer), it is said to be in the promiscuous mode.

Propagation of Risk - Spreading of risk in a network when a system with an accepted level of risk is connected to that network.

Protocol - Agreed-upon methods of communications used by computers. A specification that describes the rules and procedures that products should follow to perform activities on a network, such as transmitting data. If they use the same protocols, products from different vendors can communicate on the same network.

Prowler - A daemon that is run periodically to seek out and erase core files, truncate administrative log files, nuke lost+found directories, and otherwise clean up.

Proxy - A firewall mechanism that replaces the IP address of a host on the internal (protected) network with its own IP address for all traffic passing through it. A software agent that acts on behalf of a user, typical proxies accept a connection from a user, makes a decision as to whether or not the user or client IP address is permitted to use the proxy, perhaps does additional authentication, and then completes a connection on behalf of the user to a remote destination. See firewall.

ps - A UNIX command to display all currently active processes.

Pseudo Internet Address - Typically an ISP provides Internet service to a user customer. The ISP owns the IP address and routes any Internet mail and other data services belonging to the user to the user's gateway or computer system.

Public Data Network - A network established and operated by a telecommunication administration or a recognized private operating agency for the specific purpose of providing data transmission services for the public.

Public Domain Software - Software acquired, often at no charge. Such software commonly does not have security protection features and is more susceptible to viruses.

Public Switched Network - Any common carrier network that provides circuit switching among public users.

Purge - Removal of sensitive data from an AIS at the end of a period of processing.

put - An FTP command to upload a file on the local machine onto a remote machine.

R

RAM (Random Access Memory) - Also known as read-write memory; the memory used to execute application programs.

Rainbow - Any of several series of technical manuals distinguished by cover color. The original rainbow series was the NCSC security manuals. See Orange Book.

rcp - A UNIX command to copy files to and from remote systems. See sunrpc.

RFC (Request For Comment) - One of a long-established series of Internet informational documents and standards.

RFS (Remote File Service) - One of the many distributed file system network protocols that allow one computer to use the files and peripherals of another as if they were local. Developed by AT&T and adopted by other vendors as a part of UNIX V. See SMB.

Read-only - A file, program, directory, disk, file system, etc. that is readable by authorized users but cannot be edited due to not having write privileges.

Repeater - A device that amplifies and regenerates signals so they can travel on additional cable segments.

Replicator - Any program that acts to produce copies of itself; a program, a worm, a fork bomb, or virus. It is even claimed by some that UNIX and C are the symbiotic halves of an extremely successful replicator.

Resident Memory - Section of the central processing unit that, during processing, holds program instructions, input data, calculation results, and data to be output.

Residual Risk - The portion of risk that remains after security measures have been applied.

Response Time - The time a system or functional unit takes to react to a given input.

Retro-Virus - A retro-virus is a virus that waits until all possible backup media are infected too, so that it is not possible to restore the system to an uninfected state.

rcxd - This UNIX command is the Sun RPC server for remote program execution. This daemon is targeted by inetd whenever a remote execution request is made. See rpc.

Risk Assessment - A study of the vulnerabilities, threats, likelihood, loss or impact, and theoretical effectiveness of security measures. The process of evaluating threats and vulnerabilities, known and postulated, to determine expected loss and establish the degree of acceptability to system operations.

Risk Management - The total process to identify, control, and minimize the impact of uncertain events. The objective of the risk management program is to reduce risk, obtain, and maintain DAA approval.

rlogin - UNIX command to remotely login to another machine. User must have a user account on the remote machine.

rm - A UNIX command to remove files from a directory. The "rm *" will remove all files from a subdirectory. The most dangerous is "rm * -r" which will recursively remove files and entire subdirectories without warning.

Robust – A term used when a system has demonstrated the ability to recover gracefully from a whole range of exceptional inputs and situations in a given environment. One step below bulletproof.

ROM (Read-Only Memory) - Memory containing preloaded programs that cannot be rewritten or changed by the CPU.

Root - By extension, the privileged system-maintenance login on any OS. The superuser account with user name "root" that ignores permission bits, user number 0 on a UNIX system. The top node of a system directory structure.

Rootkit - A hacker security tool that captures passwords and message traffic to and from a computer. A collection of tools that allows an attack hacker to provide a backdoor into a system, collect information on other systems on the network, mask the fact that the system is compromised, and much more. Rootkit is a classic example of Trojan Horse software. Rootkit is widely distributed for a wide range of operating systems.

Router - An interconnection device that is similar to a bridge but serves packets or frames containing certain protocols. Routers link LANs at the network layer.

Routing Control - The application of rules during the process of routing so as to choose or avoid specific networks, links, or relays.

rpc - This is a set of library routines for performing remote procedure calls from one machine to other machines across a network.

rsh - A UNIX command that starts a remote shell on a remote machine, executing a command on the remote machine. On some systems, rsh refers to the restricted shell.

rwho - A UNIX command to determine who is logged into a remote machine.

S

Samurai - A hacker who hires out for legal cracking jobs, snooping for factions in corporate political fights, lawyers pursuing privacy rights and First Amendment cases, and other parties with legitimate reasons to need an electronic locksmith.

SATAN (Security Administrator Tool for Analyzing Networks) - A tool for remotely probing and identifying the vulnerabilities of systems on Internet Protocol networks. A powerful freeware program which helps to identify system security weaknesses.

Screened subunit - A subunit behind a screening router. The degree to which the subunit may be accessed depends on the screening rules in the router.

SCSI (small computer systems interface) - A PC bus interface standard that defines standard physical and electrical connections for devices.

SCSI device driver - A program that enables a computer to communicate with SCSI peripheral devices. Each kind of device requires a different driver.

Search-And-Destroy Mode - Hackerism for a noninteractive search-and-replace facility in an editor, so called because an incautiously chosen match pattern can cause infinite damage.

Secure Network Server - A device that acts as a gateway between a protected enclave and the outside world.

Secure Operating System - An operating system that effectively controls hardware and software functions in order to provide the level of protection appropriate to the value of the data and resources managed by the operating system.

Secure Shell - An completely encrypted shell connection between two machines protected by a super long passphrase.

Secure System Operations - Resident software that controls hardware and other software functions in an AIS to provide a level of protection or security appropriate to the classification, sensitivity, and/or criticality of the data and resources that it manages.

Security - The condition achieved when designated information, material, personnel, activities, and installations are protected against espionage, sabotage, subversion, and terrorism, as well as against loss or unauthorized disclosure.

Security Audit - A search through a computer system for security problems and vulnerabilities.

Security Breach - A violation of controls of a particular information system such that information assets or system components are unduly exposed.

Security Countermeasures - Countermeasures that are aimed at specific threats and vulnerabilities or involve more active techniques as well as activities traditionally perceived as security.

Security Incident - Any act or circumstance that involves classified information that deviates from the requirements of governing security publications, e.g., compromise, possible compromise, inadvertent disclosure, and deviation. An event involving classified information in which there is a deviation from the requirements of the governing security regulations.

Security Policy - The set of laws, rules, and practices that regulate how an organization manages, protects, and distributes sensitive information.

Security Threat - The technical and operational capability of an adversary to detect and exploit vulnerabilities.

Security Violation - An instance in which a user or other person circumvents or defeats the controls of a system to obtain unauthorized access to information contained therein or to system resources.

sed - Another UNIX editor. This editor can edit one or more files automatically simplifies edits with great pattern matching, input/output processing, and branching commands.

Sendmail - The standard UNIX mail agent. A UNIX mail program with lots of known vulnerabilities to hack into a computer system. It is very flexible, but has very difficult configuration syntax and has had numerous security bugs, because it is a large program, which needs to run with root privileges.

Server - A system that provides network service such as disk storage and file transfer, or a program that provides such a service. A kind of daemon which performs a service for the requester, which often runs on a computer other than the one on which the server runs.

Session - An activity for a period of time, the activity is accessed to a computer/network resource by a user. A period of time from log on to termination log off.

S-HTTP - Secure Hyper Text Transport Protocol.

Shareware - Software freely distributed with the understanding that users will voluntarily pay for it if they continue to use it after a short trial period. Synonymous with freeware.

Shell - A computer command interpreter used to pass commands to an operating system. In UNIX there are several programming shells, i.e., Korn shell, Bourne shell, or CShell.

shutdown - A UNIX command used to shutdown the system; has many options.

Single Sign-On - A method by which a user must identify him/herself and present their credentials only once to a system.

SMB (Server Message Block) - A distributed file system network protocol that allows one computer to use the files and peripherals of another as if they were local. Developed by Microsoft and adopted by IBM and other vendors.

SMTP (Simple Mail Transfer Protocol) - A protocol that describes an e-mail system with both host and user sections. Usually used for UNIX.

Snarf - To grab a large document or file for the purpose of using it with or without the author's permission.

Sneaker - An individual hired to break into places in order to test their security, analogous to tiger team.

Sniffer - A program to capture data across a computer network. Used by hackers to capture userID names and passwords. Software tool that audits and identifies network traffic packets.

SNMP (Single Network Management Protocol) - A structure for formatting messages and for transmitting information between reporting devices and data-collection programs, developed in part by DoD, Industry, and Academia as part of the TCP/IP protocol suite.

Social Engineering - An attack based on deceiving users or administrators at the target site. Social engineering attacks are typically carried out by telephoning users or operators and pretending to be an authorized user to attempt to gain illicit access to the systems.

Spam - To crash a program by overrunning a fixed-site buffer with excessively large input data. Also, to cause a person or newsgroup to be flooded with irrelevant or inappropriate messages.

SPI (Secure Profile Inspector) - A network monitoring tool for UNIX, developed by the Department of Energy.

Spoofing - Pretending to be someone else. The deliberate inducement of a user or a resource to take an incorrect action. Attempt to gain access to an AIS by pretending to be an authorized user. Impersonating, masquerading, and mimicking are forms of spoofing.

Spooling - To send files to some device or program that queues them up and does something useful with them later. The use of auxiliary storage as buffer storage to reduce processing delays when transferring data between peripheral equipment and the processors of a computer. Spooling directory, spooling space.

ssh - A UNIX command which allows logging into a remote machine and for executing commands in a remote machine. It is intended to replace rlogin and rsh and provide secure encrypted communications between two untrusted hosts over an insecure network.

SSL - Secure Sockets Layer.

su - A UNIX command requesting a change to superuser mode. See Superuser.

sunrpc - A Sun/UNIX command to copy files to and from remote systems. See rpc.

Superuser - The account with all privileges. Root account. System Administrator or System Administrator-level privileges. A user who is granted special privileges if the correct password is supplied when logging in as root or using the su command. This usage has spread to non-UNIX environments; the superuser is any account with all wheel bits on. Special user with user name root and useridentifier 0, who can perform in control of processes, devices, networks, and file systems.

Surf - To traverse the Internet in search of interesting stuff.

swap space - A disk space, partition, or file used to temporarily hold the contents of a memory area until it can be loaded back into memory.

SYN Flood - When the SYN queue is flooded, no new connections can be opened. See denial of service.

Synchronous - Refers to a transmission system in which characters are synchronized by the transmission of initial sync characters and a common clock signal. No stop or start bits are used.

SysAdmin - Common contraction of system administrator.

SysOp - The System Operator. In the BBS world, the operator of a bulletin-board system.

System - A computer with a keyboard and terminal. A system can have either local or remote disks and may have additional peripheral devices.

System Administrator - A computer operations person whose primary function is to operate on all equipment associated with the computer system.

T

T1 - A 1.544 megabit per second communications circuit provided by long-distance communications carriers for voice or data transmission. T1 lines are typically divided into 24 separate 64 kilobit channels.

Talk mode - A feature supported by UNIX and some other operating systems that allows two or more logged-in users to set up a real-time on-line conversation. Automatically splits the user screens into windows per number of users.

tar - A UNIX command to archive files, often on backup tapes. Follow-on tar commands can append files to existing “tar-ed” files.

TCP/IP (Transmission Control Protocol/Internet Protocol) - This is the wide area networking protocol that makes the Internet work. A set of communications protocols that has evolved since the late 1970s when it was first developed by the DoD. Because programs supporting these protocols are available on so many different computer system, they have become an excellent way to connect different types of systems over networks.

tcpwrapper - A software tool for security which provides additional network logging and restricts service access to authorized hosts by service.

TELNET - A terminal-emulation protocol. Software supporting TELNET usually comes as a utility in a TCP/IP package. Included as a standard feature of UNIX. TELNET also supports an input mode, where one can enter commands directly on the remote system.

Tentacle - An artificial identity created in cyberspace for malicious and deceptive purposes. The implication is that a single person may have multiple tentacles.

Terminal Hi-jacking - Allows an attacker on a certain machine to control any terminal session that is in progress. An attack hacker can send and receive terminal i/o while a user is on the terminal.

tftp - The UNIX command for the user interface to the Internet TFTP, which allows users to transfer files to and from a remote machine.

TFTP (Trivial File Transfer Protocol) - A simplified version of FTP that transfers files but does not provide password protection or user-directory capability. It is associated with the TCP/IP family of protocols.

Threat - The means through which the ability or intent of a threat agent to adversely affect an automated system, facility, or operation can be manifest. A potential violation of security.

Threat Agent - Methods and techniques used to exploit a vulnerability in an information system, operation, or facility.

Threat Assessment - Process of formally evaluating the degree of threat to an information system and describing the nature of the threat.

Threat Event - A specific type of threat event as often specified in a risk analysis procedure.

Throughput - The number of bits, characters, or blocks that can pass through a data communication system when the system is working at saturation.

Tiger - A software tool which scans for system weaknesses.

Tiger team - Government and industry-sponsored teams of computer experts who attempt to break down the defenses of computer systems in an effort to uncover, and eventually patch, security holes.

Time Bomb - A logic bomb that is triggered by reaching some preset time, either once or periodically. A variant of the Trojan Horse in which a malicious code is inserted to be triggered later.

Time-Out - A network parameter related to an enforced event designed to occur at the conclusion of a predetermined elapsed time.

Time-Sharing - A mode of operation that provides for the interleaving of two or more independent processes on one function unit.

Tinkerbell Program - A monitoring program used to scan incoming network calls and generate alerts when calls are received from particular sites or when logins are attempted using certain IDs.

Topology - The map or plan of the network. The physical topology describes how the wires or cables are laid out and the logical or electrical topology describes how the messages flow.

touch - A UNIX command to change a file's access time and modification time to the current date. If trying to change the date for a file that does not exist, touch will create a new file under that name. This is one quick way to determine if a directory structure can be written to.

Trace Packet - In a packet-switching network, a unique packet that causes a report of each stage of its progress to be sent to the network control center from each visited system element.

traceroute - An operation of sending trace packets for determining information and traces the route of UDP packets for the local host to a remote host. Normally traceroute displays the time and location of the route taken to reach its destination computer.

Traffic - The information moved over a communication channel.

Transceiver - A communicating device capable of transmitting and receiving.

Trap Door - A hidden software or hardware mechanism used to circumvent security control. See Back door.

Tripwire - A software tool for security. Basically, it works with a database that maintains information about the byte count of files. If the byte count has changed, it will notify the system security manager.

Trojan Horse - An apparently useful and innocent program containing additional hidden codes which allows the unauthorized collection, exploitation, falsification, or destruction of data.

Troll - An online message whose purpose is to attract responses and make the responders look stupid. People who troll want to make you waste your time responding to their pointless statements.

True-Hacker - One who exemplifies the primary values of hacker culture, especially in competence and helpfulness to other hackers.

Trust/Trusted - Confidence that an entity, to which trust is applied, will perform in a way that will not prejudice the security of the user of the system of which that entity is a part. (e.g., agent, a computer, a network, a database, etc.).

TTY Watcher - A hacker tool that allows hackers with even a small amount of skill to hi-jack terminals. It has a GUI interface and reported to be loads of fun. See Terminal Hi-jacking.

Twisted Pair Wiring - Cable comprised of two wires twisted together at six turns per inch to provide electrical self-shielding.

Type 1 - Cryptographic systems, products/algorithms approved for the protection of classified U.S. Government information.

Type 2 - Cryptographic systems, products/algorithms approved for the protection of sensitive unclassified U.S. Government information.

U

ufs - UNIX file system.

UDP (User Datagram Protocol) - A TCP/IP protocol describing how messages reach application programs within a destination computer. This protocol is normally bundled with IP-layer software.

UID number - The user identification number assigned to each login name. UID numbers are used by the system to identify, by number, the owners of files, and directories.

umask - A UNIX command to create or return the current value of the file-creation mask, which determines default values for new files.

Unapproved Software - All software that has not been formally identified, evaluated, and examined by competent personnel to ensure that the software performs to exact specifications.

Unauthorized/unprivileged User - Any person who interacts without authorization directly with a computer system.

uncompress - A UNIX command to uncompress a compressed file, usually with a name ending in .Z.

UNIX - Not an acronym. A computer operating system. UNIX has become the most widely used multitasking, multiuser, time-sharing, general purpose operating system. Developed by AT&T and has enjoyed popularity among engineering and technical professionals. UNIX is finding new uses as the basis of the file-server operating systems for networks of PCs. Reimplemented in the C programming language made it the first source-portable operating system.

Unprotected - A computer system that is not regulated with concern to the control of access between subjects and objects. See copy protected.

Upload - To transfer programs or data over a digital communications link from a smaller or peripheral client system to a larger or central host.

unzip - Command to return a "zip" compressed file back to its original uncompressed state.

URL - Uniform Resource Locator.

user account - An account set up for an individual user, (in UNIX) in the Passwd database that specifies the user's login name, UID, GID, login directory, and login shell.

User Profile - Patterns of a user's activity, which can detect changes in normal routines.

UUCP (UNIX-to-UNIX Copy Program) - A standard UNIX utility used for information exchange between two UNIX nodes. The store and forward network consisting of all the world's connected UNIX machines and others running some clone of the UNIX-to-UNIX-Copy software.

uudecode - A UNIX command that reads a file converted by uuencode and restores it to its original form.

uuencode - A UNIX command that converts a binary file to an encoded form that can be sent with the mailx command. This encoded file is in ASCII form.

uulog - A UNIX command that keeps a log of uucp file transfers to and from a specified system.

V

Vaccines - Program that injects itself into an executable program to perform a signature check and warns if there have been any changes. See Anti-Virus Program.

Veronica - A Gopher based resource that you can use to search gopherspace for all the menu items that contain specified words. See Jughead, Archie, Gopher.

vi - A text editor considered the de facto standard UNIX editor and a nearly undisputed hacker favorite until the rise of EMACS (about 1984). Considered powerful but has a rudimentary interface and tends to frustrate new users to no end.

Virtual memory - A memory management technique that is used by the operating system for programs that require more space in memory than can be allotted to them. The UNIX kernel moves only pages of the program currently needed into memory; unneeded pages remain on the disk.

Virtual network perimeter - A network that appears to be single protected network behind firewalls, which actually encompasses encrypted virtual links over untrusted networks.

Virtual Password - A password computed from a passphrase that meets the requirements of password storage, e.g., 64 bits for DES.

Virus - A variation of Trojan Horse. It is propagating with a triggering mechanism (event time) with a mission (delete files, corrupt data, send data). Often self replicating, malicious program segment that attaches itself to an application program or other executable system component and leaves no obvious signs of its presence.

Virus scanner - A software program which can search out, locate, and possibly remove a virus.

Vulnerability - Hardware, firmware, or software flow that leaves a computer processing system open for potential exploitation. A weakness in automated system security procedures, administrative controls, physical layout, internal controls, and so forth, that could be exploited by a threat to gain unauthorized access to information or disrupt critical processing.

Vulnerability Analysis - Systematic examination of an information system or product to determine the adequacy of security measures, to identify security deficiencies, to provide data from which to predict the effectiveness of proposed security measures, and to confirm the adequacy of such measures after implementation.

W

WAIS (Wide Area Information Service) - An Internet service that allows you to search a large number of specially indexed databases.

wall - A UNIX command to write a message to everyone's display screen of the immediate host LAN.

Wallpaper - A file containing a listing (e.g., assembly listing) or transcript, especially a file containing a transcript of all or part of a login session. Also used to name the background picture of windowing software.

WAN (Wide Area Network) - A type of network that connects computers over areas potentially as wide as the entire world.

Wannabee - A would-be hacker.

War Dialer - A cracking tool; a program that calls a given list or range of numbers and records those which answer with handshake tones (and so might be entry points to computer or telecommunications systems).

Warez - Widely used in cracker subcultures to denote cracked versions of commercial software, i.e., versions from which copy protection has been stripped. Hackers recognize this term but do not use it themselves.

Web Browser - A software application which allows access to Web sites on the World Wide Web. Examples are Netscape, Mosaic, Microsoft Explorer, and Apache.

Webmaster - The person at a site providing World Wide Web information who is responsible for maintaining their web site.

Web Server - A computer which feeds documents and media to the browsers that request them.

Web Site - A computer running a web server with a valid URL, which is accessible by a web browser.

Wedged - To be stuck; incapable of proceeding without help. The system or software is trying to do something but cannot make progress. It may be capable of doing a few things, but not be fully operational.

Wheel - A person who has an active wheel bit. The traditional name of security group zero, to which the major system-internal users like "root" belong. Only the big wheels have access to the wheel group.

Wheel bit - A privilege bit that allows the possessor to perform some restricted operation on a timesharing system, such as read or write any file on the system regardless of protections, change or look at any address in the running monitor, crash or reload the system, and kill or create jobs and user accounts.

who - A UNIX command to find out who is on the local machine. See rwho.

Wide Area Network (WAN) - A physical or logical network that provides capabilities for a number of independent devices to communicate with each other over a common transmission-interconnected topology in geographic areas larger than those served by local area networks. A network which ranges from a few kilometers to thousands of kilometers and handles transmission speeds from a few kilobits/second to megabits/second.

Wideband - Refers to a channel or transmission medium capable of passing more frequencies than a standard 3kHz voice channel.

Window - A portion of a display surface in which display images pertaining to a particular application can be presented. Different applications can be displayed simultaneously in different windows.

Wiring Hub - A cabinet, usually mounted in a wiring closet, that holds connection modules for various kinds of cabling. The hub contains electronic circuits that re-time and repeat the signals on the cable. The hub may also contain a microprocessor board that monitors and reports on network activity.

Wizard - A person who knows how a complex piece of software or hardware works. Especially someone who can find and fix bugs quickly in an emergency.

World Wide Web - A hypertext based tool that allows one to retrieve and display data based on keyword searches. The power of the web lies with the idea of hypertext: data that contains links to other data. See Hypertext.

Work Station - For automated systems; a configuration of input, output, display, and processing equipment that provides an operator interface to a system, such as a central computer, communication, or control system.

Worm - A program or executable code module which resides in distributed systems or networks. It will replicate itself, if necessary, in order to exercise as much of the system's resources as possible for its own processing. Such resources may take the form of CPU time, I/O channels, or system memory. It will replicate itself from machine to machine across network connections, often clogging networks and computer systems as it spreads.

Write - A UNIX command to send a note or message to another users display screen. See Wall.

X

X - A windowing software application and widely used on UNIX system.

X Window - A network-based windowing system that provides a programmatic interface for graphic window displays. X Window permits graphics produces on one networked workstation to be displayed on another.

Y

Yellow Pages - A network information service; renamed to Yellow Pages. See NIS.

yppasswd - The UNIX command for changing the network password associated with the username in the Network Information Service database. See nispasswd.

Z

Zero, Zero-fill, Zero-out - To fill unused storage locations in an IS with zeros. Normally done to erase or remove information on magnetic media.

Zip - To create a compressed archive from a file or group of files; originally from using PKWare's PKZIP program.

Special thanks to Byron B. Thatcher, Major, USAF, AFIWC/EACA for supplying the information used in this section.

Appendix B – Computer Emergency Response Teams

The following is a list of Computer Emergency Response Teams (CERTs). These organizations have been formed to help organizations and companies deal with computer security-related issues. These types of issues usually involve criminal hacker attacks, known security vulnerabilities in systems and applications, viruses, and Trojan Horse programs. A company or organization can contact these CERTs to receive assistance.

PLEASE ENSURE THAT YOUR COMPANY OR ORGANIZATION BELONGS TO THE CONSTITUENCY LISTED FOR EACH CERT. THE CERTs WILL ONLY ASSIST THOSE ENTITIES THAT ARE WITHIN THEIR CONSTITUENCIES.

The form listed in "Appendix B – Computer Security Incident Report Form" should be completed prior to contacting the appropriate CERT. This will help the CERT to be more effective and efficient and to begin the critical documentation process. While CERTs are valuable, they should be used prudently.

*** Please note that some of the following information may have changed since this information was published.**

Constituency: Air Force - U.S. Air Force
Response Team: AFCERT
E-mail: afcert@afcert.csap.af.mil

Constituency: Department of Defense (Services and Command)
Response Team: DoD's ASSIST (Automated Systems Security Incident Support Team)
E-mail: dod-cert@ddn-conus.ddn.mil

Constituency: Digital Equipment Corporation and Customers
Response Team: SSRT (Software Security Response Team)
E-mail: rich.boren@cxo.mts.dec.com

Constituency: Energy — U.S. Dept. of Energy sites and Energy Sciences Network (ESnet)
Response Team: CIAC (Computer Incident Advisory Capability)
E-mail: ciac@llnl.gov

Constituency: Germany — Southern Area
Response Team: Micro-BIT Virus Center
E-mail: ry15@rz.uni-karlsruhe.de

Constituency: Internet
Response Team: CERT/CC (Computer Emergency Response Team/Coordination Center)
E-mail: cert@cert.org

Constituency: Motorola
Response Team: Motorola Computer Emergency Response Team (MCERT)
E-mail: mcert@mot.com

Constituency: MILNET
Response Team: DDN (Defense Data Network)
E-mail: scc@nic.ddn.mil

Constituency: NASA and NASA-Sponsored Wide-Area TCP/IP and DECnet Systems
Response Team: NASA Science Internet (NSI) Security Office
E-mail: security@nsinic.gsfc.nasa.gov
Secondary E-mail: tencati@nsi-security.gsfc.nasa.gov

Constituency: NASA — NASA Ames Research Center
Response Team: NASA Ames
E-mail: yee@ames.arc.nasa.gov

Constituency: Navy — U.S. Department of the Navy
Response Team: NAVCIRT (Naval Computer Incident Response Team)
E-mail: navcirt@dockmaster.ncsc.mil

Constituency: Penn State — The Pennsylvania State University
Response Team: Penn State
E-mail: lambert@cs.psu.edu

Constituency: Purdue University
Response Team: PCERT
E-mail: pcert@cs.purdue.edu

Constituency: SPAN-France
Response Team: SPAN France
E-mail: harvey%meudon.dnet@east.gsfc.nasa.gov

Constituency: Sprint — U.S. Sprint
Response Team: Sprint DNSU
E-mail: /PN=DATANETWORK.SECURITY/O=US.SPRINT/ADMD=TELEMAIL/C=US/@sprint.com
E-mail: /PN=CORPINFO.SECURITY/O=US.SPRINT/ADMD=TELEMAIL/C=US/@sprint.com

Constituency: Customers of Sun Microsystems
Response Team: Sun Microsystem's Customer Warning System (CWS)
E-mail: security-alert@sun.com

Constituency: SURFnet connect sites, Netherlands
Response Team: SURFnet Computer Emergency Response Team
E-mail: cert-nl@surfnet.nl

Constituency: TRW Network Area and System Administrators
Response Team: TRW's CERCUS (Computer Emergency Response Committee for Unclassified Systems)
E-mail: cercus@gumby.dsd.trw.com

Constituency: UK — all government departments and agencies
Response Team: CCTA IT Security & Infrastructure Group
E-mail: carr@csrc.ncsl.nist.gov

Constituency: Unisys Internal and External Users
Response Team: UCERT
E-mail: garb@dockmaster.ncsc.mil

Appendix C – Computer Security Incident Report Form

(Based on a form from NORDUnet CERT)

This is an example of a form used for reporting computer security incidents to organizations like Computer Emergency Response Team (CERT). There are a number of CERT-like organizations throughout the world. It is important to identify the one which will be able to assist your company in the event of security incident (see Appendix A). By preparing the following information before contacting them, it will save valuable time and effort.

1. Incident number (assigned by the responding organizations):

2. Reporting site information:

    Organizational Name (e.g., NORDUnet CERT):

    Domain Name (e.g., nordu.net):

3. Your contact information:

    Name:

    E-mail address:

    Telephone number:

    Fax number (optional):

    Pager number (optional):

    Home telephone number (for internal use only):

4. Additional contact information (if available):

    Name:

    E-mail address:

    Telephone number:

    Fax number (optional):

    Pager number (optional):

    Home telephone number (for NDN-CERT internal use only):

5. Compromised host(s) at your site (one entry per host please):

   Hostname:

   IP address:

   Vendor:

   Hardware:

   OS:

   Version:

   Security patches applied:

6. Please list the other sites compromised that you have notified and the contact information for each site (one entry per site please):

   Hostname:

   IP address:

   Contact name:

   E-mail address:

   Telephone number:

   Fax number (optional):

   Pager number (optional):

   Home telephone number (optional, for internal use only):

7. Please list the other sites compromised that you have not yet notified (one entry per site):

   Hostname:

   IP address:

   Contact name:

   E-mail address:

   Telephone number:

   Fax number (optional):

   Pager number (optional):

   Home telephone number (optional, for internal use only):

8. Would you be willing to contact these sites if you had the relevant contact information (Yes/No)?

   Or, can your contact information be given to these sites when they are contacted (Yes/No)?

9. Incident category (Yes/No):

   Probe

   Prank

   Mail Spoofing

   Break-in

   Installed Trojan Horse

   Intruder gained root access

   NIS (yellow pages) attack

   NFS attack

   TFTP attack

   FTP attack

   Telnet attack

   Rlogin or rsh attack

10. Product vulnerability:

    Worm

    Virus

    Application

    Operating System

    Other (please specify)

11. Are you currently using any of the following? (Yes/No/Periodically):

    COPS (The Computer Oracle and Password System)

    TCP access control using packet filtering

    Host access control via modified daemons or wrappers

    Crack

    Tripwire

Proactive password checkers (e.g., npasswd)

Shadow passwords

Other (please specify)

12. Please specify any other incident response team(s) you have contacted:

Team:

Contact name:

E-mail address:

Telephone number:

Fax number (optional):

Pager number (optional):

Home telephone number (optional, for internal use only):

If you have not contacted another incident response team could they be given your contact information (Yes/No)?

13. Please specify any police authority(ies) or government agencies you have contacted:

Contact name:

E-mail address:

Telephone number:

Fax number (optional):

Pager number (optional):

Home telephone number (optional, for internal use only):

If you have not contacted any police authority could we give them your contact information if necessary (Yes/No)?

14. Detailed description of incident (e.g., method of intrusion, etc):

15. What kind of assistance would you like?

16. Append any log information or altered directory listings to a safe area for investigative purposes.

Appendix D — Common File Extensions

.$$f	Database File, OS/2
.$$p	Notes, OS/2
.$$s	Spreadsheet File, Pmspread, OS/2
.d	Planner File, OS/2
.$db	Temporary File, Dbase IV
.$ed	Temporary Editor File, MS C
.$vm	Temporary File From Virtual Managers, MS Windows
.?q?	Squeeze
.~	Temporary File, MS Windows and others
.~mn	Menu Back-Up, Norton Commander
.075	75x75 dpi Screen Characters, Ventura Publisher
.085	85x85 dpi Screen Characters, Ventura Publisher
.091	91x91 dpi Screen Characters, Ventura Publisher
.096	96x96 dpi Screen Characters, Ventura Publisher
.0b	Printer Font Linedraw Enhanced Characters, Pagemaker
.15u	Printer Font, Pagemaker
.1st	ASCII Text, program information
.286	Driver Standard Mode, MS Windows
.2gr	Screen Grabber EGA/VGA Mode, MS Windows
.301	Fax Data File, Super Fax 2000
.386	Driver 386 Enhanced Mode, MS Windows
.3ds	Graphic, 3D Studio
.3fx	3-D Effect, Corel Chart
.3gr	Screen Grabber SVGA/XVGA Mode, MS Windows
.4sw	Swap File, 4DOS
.4th	Forth Program File, FORTH CMP
.8	a86 Assembler Program File
.8m	Printer Font, Math 8 enhanced characters, Pagemaker
.8u	Printer Font, Roman 8 enhanced characters, Pagemaker
.a	ADA program file
.a	Library, UNIX
.a_t	A-TRAIN
.a11	Graphic
.ab6	Data File, AB Stat
.ab8	Data File, AB Stat
.abc	Flowchart, ABC Flowcharter 1.x
.abk	Automatic Back-Up, Corel Draw
.abs	Information, Abstracts
.acm	Driver, Audio Compression Manager
.acr	Acr-Nema Medical Image File
.act	Actor Program File
.act	Foxdoc Action Diagram, FoxPro
.ad	Screen Saver File, After Dark
.ada	ADA Program File
.adb	ADA Package Body
.add	Device Driver, Adapter device driver
.adf	Admin Config File
.adi	Plotter File, AutoCAD
.adl	MCA Adapter Description Library, QEMM
.adm	Screen Saver Module, After Dark
.adn	Add-In, Lotus 1-2-3
.adr	Random Screen Saver Module, After Dark
.ads	ADA Package Specification
.adx	Index File, Approach
.af2	Flowchart, ABC Flowcharter 2.0
.af3	Flowchart, ABC Flowcharter 3.0
.afl	Font, Lotus 1-2-3
.afm	Font, Outline font Metric
.afp	Symbol Palette, ABC Flowcharter
.aft	Template, ABC Flowcharter
.afw	Work Area, ABC Flowcharter
.ai	Vector Graphic, Adobe Illustrator
.aif	Setup Information, ADW Knowledge Ware
.aif(c	Sound File, Apple, SGI
.aif(f	Sound File, Apple, SGI
.alo	APL File Transfer Format
.ais	Graphic, Array of Intensity Samples
.all	Printer Information, WordPerfect For Windows
.all	Symbols And Characters, Arts & Letters
.alt	Menu File, WordPerfect
.ani	Animation
.anm	Animation, Deluxe Paint
.ann	Help Annotation, MS Windows
.ans	Text And Layout, MS Word For Windows
.ant	Sim Ant
.apc	Printer Driver Characters, Lotus 1-2-3
.apd	Printer Driver, Lotus 1-2-3
.apf	Printer Driver Fonts, Lotus 1-2-3
.api	Printer Driver Information, Lotus 1-2-3
.app	Add-In Application File, Symphony
.app	Application, DR-DOS/Next Step
.app	Relational Application, FoxPro

.arc	Archive File, MS Schedule
.arc	Compressed Archive File, ARC
.arc	DOS Compression
.arj	Compressed
.arj	Compressed Archive File, ARJ
.art	First Publisher Clip Art File
.art	Raster Graphic, Atari
.asc	ASCII Text File
.asc	Image File
.asd	Automatic Back-Up, MS Word For Windows
.asd	Screen Driver, Lotus 1-2-3
.asf	Data File, StatGraphics
.asf	Screen Font, Lotus 1-2-3
.ash	Assembler Header File, Turbo Assembler
.asi	Assembler Include File, Turbo C/Borland C++
.asm	Assembler Program File
.aso	Assembler Object, Turbo Assembler
.asp	Aspect Program File, ProComm Plus
.asp	Association of Shareware Professionals
.au	Audio File, Sun Microsystems
.au	Music File
.aux	Help Dictionary File, TEX/Chiwriter
.avi	Video, MS Windows
.avi	Animations File, Audio Video Interleaved, Video For Windows
.avr	Sound File
.avs	Animation
.awk	Awk Script/Program
.aws	StatGraphics
.b	Batch List, Applause
.b30	Printer Font, JLaser - Cor file, Ventura Publisher
.b44	Network Plan, BETA 44
.b8	Raw Graphic
.bac	Binary Back-Up
.bak	Back-Up
.bas	Basic File
.bas	Basic Program File
.bat	Batch File, DOS
.bb	Database Back-Up, Papyrus
.bbm	Image File
.bck	VMS Back-Up
.bco	Outline Font
.bgi	Borland Graphics Interface Device Driver
.bib	Bibliography, ASCII
.big	Chinese, old version
.big5	Chinese, old version
.bin	Binary File
.bio	OS/2 Bios
.bit	Bitmap X11
.bk!	BACK-UP, WordPerfect For Windows
.bk1	Back-Up, WordPerfect For Windows
.bk2	Back-Up, WordPerfect For Windows
.bk3	Back-Up, WordPerfect For Windows
.bk4	Back-Up, WordPerfect For Windows
.bk5	Back-Up, WordPerfect For Windows
.bk6	Back-Up, WordPerfect For Windows
.bk7	Back-Up, WordPerfect For Windows
.bk8	Back-Up, WordPerfect For Windows
.bk9	Back-Up, WordPerfect For Windows
.bki	Back-Up Index, IBM Bookmanager
.bkp	Back-Up, Write
.bks	Back-Up Main, Bookshelf, IBM Bookmanager
.bld	Basic File In BLoad Format
.blk	Temporary File, WordPerfect For Windows
.bm	Graphic, bitmap
.bmf	Image File
.bmk	Bookmark For Help File, Bookmarks, MS Windows
.bmp	Picture File, Bitmap
.bmp	Bitmap Graphic, MS Windows, PC Paintbrush
.bnk	Adlib Instrument Bank
.boo	Image File
.boo	Book, IBM Bookmanager
.boo	Compressed Archive File, BOO
.bpt	Bitmap Master File, Corel Draw
.btm	Batch To Memory Batch File, 4DOS
.btn	Button File, ButtonWare
.bup	Back-Up
.bv1	Back-Up File, WordPerfect For Windows
.bv2	Back-Up File, WordPerfect For Windows
.bv3	Back-Up File, WordPerfect For Windows
.bv4	Back-Up File, WordPerfect For Windows
.bv5	Back-Up File, WordPerfect For Windows
.bv6	Back-Up File, WordPerfect For Windows
.bv7	Back-Up File, WordPerfect For Windows
.bv8	Back-Up File, WordPerfect For Windows
.bv9	Back-Up File, WordPerfect For Windows

.byu	Movie Byu File
.c	C Program File
.c00	Print File, Ventura Publisher
.c86	C86 C Program File
.ca7	Job File, BETA 44
.cal	Calendar File, MS Windows
.cal	Spreadsheet File, SuperCALC
.cap	Caption, Ventura Publisher
.cap	Session Capture File, Telix
.cat	Catalog, Dbase IV
.cbl	Cobol Program File
.cbt	Computer Based Training
.cc	C++ Program File
.cch	Chart, Corel Chart
.cdb	Database, Turbo C Utilities
.cdr	Vector Graphic, Corel Draw
.cel	Graphic, AutoDesk Animator
.cfg	Configuration
.cga	CGA Screen Characters, Ventura Publisher
.cgm	Computer Graphics Metafile
.cgm	Vector Graphic, Computer Graphics Metafile
.ch	Header File, Clipper 5
.ch3	Chart- File, Harvard Graphics
.chi	Document, CHIWriter
.chk	Lost Cluster Data From Chkdsk, DOS
.chk	Temporary File, WordPerfect For Windows
.chp	Chapter, Ventura Publisher
.chr	Characters, Turbo C/Turbo Pascal
.cht	Chart, Harvard Graphics 2.0
.cht	Interface For Chartmaster, dBase
.cif	Chapter Information, Ventura Publisher
.cif	Graphic, Caltech Intermediate Format
.cix	Database Index, Turbo C Utilities
.cl	Common Lisp Program File
.clp	Picture File
.clp	Clipart Graphic, Quattro Pro
.clp	Clipboard File, MS Windows
.clp	Compiler Script, Clip List, Clipper 5
.cls	C++ Class Definition
.cm	Macro
.cmd	Batch Command, OS/2
.cmd	Command File, dBase
.cmf	Music File
.cmf	Creative Music File: Sound File
.cmp	Header Postscript Printer, Corel Draw
.cmp	Users Dictionary File, MS Word
.cmy	Color Palette
.cnf	Configurations File
.cnt	Help/Index
.cnv	Data Conversion File, MS Word For Windows
.cnv	Temporary File, WordPerfect For Windows
.cob	Cobol Program File
.cod	Compiler Program Code, Fortran
.cod	Data File, MS Multiplan
.cod	Template Source File, dBase
.col	Color Palette, AutoDesk Animator
.col	Spreadsheet File, MS Multiplan
.com	Compressed
.com	Executable File
.com	VMS Command File
.com	Program
.con	Consolidation File, ADW Knowledge Ware
.cpi	Code Page Information
.cpi	Graphic, ColorLab Processed Image
.cpl	Applet: Control Panel, MS Windows
.cpp	C++ Program File
.cps	S Checksum For Anti-Virus, PCTools
.cpt	Macintosh Compressed
.cpx	Applet: Control Panel, MS Windows
.crd	Index Card File, MS Windows
.crf	Cross-Reference, MS MASM/Zortech C++
.crp	Database, Dbase IV
.csq	Queries, FoxPro
.csv	Comma Separated Values: Text File, ASCII
.ctf	Character Code Translation File, Symphony
.ctl	Setup Information
.ctx	Compressed Text
.cuf	C Utilities Form Definition, Turbo C Utilities
.cur	Image File
.cur	Cursor, MS Windows
.cut	Image File
.cut	Bitmap Graphic, DR. Halo
.cv4	Colors, Codeview
.cvt	Back-Up Converted Database, Dbase IV
.cvw	Colors, Codeview
.cxx	C++ Program File, Zortech C++
.dat	VMS Data File
.dat	Data, usually in ASCII, Registration, MS Windows/regedit
.db	Configuration, dBASE IV/dBFAST
.db	Database, Paradox
.db$	Temp Debug Information, Clarion ModuleA-2
.db2	Database, dBASE II
.db3	Database, dBASE III

.dba	Data File, Dataease
.dbd	Debug Information, Clarion ModuleA-2
.dbf	Database, dBASE/dBFAST/Foxpro
.dbg	Symbolic Debug Information, MS C/C++
.dbk	Database Back-Up, Dbase IV
.dbm	Data File, Dataease
.dbo	Compiler Program, Dbase IV
.dbs	Data File, PRODAS
.dbs	Database In Sql Windows Format
.dbs	Printer Description, Works
.dbt	Foxbase+ Style Memo, Foxpro
.dbt	Memo File For Database, dBASE IV/dBFAST
.dca	Text File, Document Content Architecture, IBM DISPlaywrite
.dcp	File Codepage, OS/2
.dcs	Image File
.dct	Dictionary File, Harvard Graphics, Symphony, CRYstal reports
.dd	Compressed Archive File Macintosh, Diskdoubler
.ddi	Diskdupe Image, Diskdupe
.ddp	Device Driver Profile File, device driver Profile, OS/2
.def	Definitions, Defaults
.del	List of Data
.dem	Demonstration
.des	Text File
.dev	Device Driver
.dfd	File Flow Diagram Graphic, Prosa
.dfm	File Flow Diagram Model, Prosa
.dfv	Print Format Template, Word
.dhp	Graphic, PIC Format, DR. Halo
.dib	Picture File, Bitmap
.dib	Device-Independent Bitmap Graphic, MS Windows
.dic	Dictionary File
.dif	Database, Visicalc
.dif	Text File, File Interchange Format
.dig	Text
.dir	Dialing Directory, Procomm Plus
.dis	Thesaurus, Corel Draw
.diz	Program Description
.dl	Animated Picture
.dl	Animations File
.dld	Lotus 1-2-3
.dlg	Image File
.dlg	Dialog Resource Script, MS Windows & SDK
.dll	Dynamic Link Library, MS Windows/OS/2

.dmo	Demo
.dmp	Dump File
.dms	Compressed
.dms	Compressed Archive File Amiga
.doc	Text
.doc	Document, Text file
.dos	Text File And DOS Specification Info
.dot	Document Template, MS Word For Windows
.dot	Lines-Definition, Corel Draw
.dox	Text File, Multimate 4.0
.dpr	Default Project, Borland C++
.drs	Display Resource, Wordperfect For Windows
.drv	Device Driver
.drw	Vector Graphic, MICROGRAFIX DESIGNER
.dsk	Project Desktop, Borland C++
.dsp	Graphicscreen Driver, DR. Halo
.dsr	Driver, Wordperfect For Windows
.dta	Data File, Turbo Pascal/PC-FILE
.dtf	Database, Q&A
.dvc	Lotus 1-2-3
.dvi	Device Independent Document, TeX
.dvp	Desqview Program Information File, Desqview
.dvp	Device Parameter File, AutoCAD
.dwc	Compressed Archive File
.dwg	Drawing Database, AutoCAD
.dwg	Image File
.dxf	ASCII Drawing Interchange
.dxf	Drawing Interchange File Format Vector Graphic, AutoCAD
.dyn	Lotus 1-2-3
.eeb	Button Bar For Equation Editor, Wordperfect For Windows
.eft	High Resolution Screen Characters, CHIWriteR
.ega	EGA Screen Characters, Ventura Publisher
.elt	Event List Text File, Prosa
.ems	Enhanced Menu System Configuration, PC Tools
.emu	Emulation, IRMA Workstation For Win
.enc	Encyclopedia, ADW Knowledge Ware
.end	Arrow Definition, Corel Draw
.eng	Dictionary File Engine, Sprint
.env	Environment, Wordperfect For Windows
.eps	Encapsulated Postscript Vector Graphic, Adobe Illustrator
.eps	Printer Font, Ventura Publisher

.eqn	Equation File, Wordperfect For Windows
.erd	Entity Relationship Diagram Graphic File, Prosa
.erm	Entity Relationship Diagram Model File, Prosa
.err	Error Log File
.esh	Extended Shell Batch File
.euc	Japanese, Kanji
.evt	Event File Scheduler, PC Tools
.ex3	Device Driver, Harvard Graphics 3.0
.exc	Exclude File From Optimization, QEMM
.exc	Rexx Program File, VM/CMS
.exe	Executable File
.exe	Self-Extracting File
.exe	Program
.ext	Extension File, Norton Commander
.f	Freeze For UNIX
.f	Compressed Archive File, FREEZE
.f	Fortran Program File
.f77	Fortran 77 Program File
.fac	Picture File
.fac	Usenix Face Graphic
.faq	Frequently Asked Questions Text File
.fax	Fax Graphic File, WINFax PRO
.fc	Dictionary File, Harvard Graphics 2.0
.fct	Catalog, Foxpro
.fd	Declarations File, MS Fortran
.fd	Field Offsets Compiler, Dataflex
.fex	Focexec, Focus
.fft	Dca/Fft Final Form Text Text File, DisplayWrite
.fi	Interface File, MS Fortran
.fif	Fractal Image Format
.fil	Files List Object File, dBASE
.fil	Overlay, Wordperfect
.fin	Print Format Text File, Perfect WriteR/MINCE
.fit	Fits Graphic
.fit(s	Image File
.fky	Macro File, Foxpro
.flb	Format Library, Papyrus
.flc	Animated Picture
.fld	Folder, CHARISMA
.fli	Animated Picture
.fli	Animation, Autodesk Animator
.fll	Library, Foxpro For Windows
.flm	Film, AutoCAD
.flt	Graphicfilter, diverse
.flx	Image File
.flx	Ordered Binder File, Dataflex
.fm1	Spreadsheet File, Lotus 1-2-3 2.x
.fm3	Device Driver, Harvard Graphics 3.0
.fm3	Spreadsheet File, Lotus 1-2-3 3.x
.fmb	File Manager Button Bar, Wordperfect For Windows
.fmo	Ordered Format File, Dbase IV
.fmt	Format File, Clipper 5/dBASE IV/Foxpro
.fmt	Style Sheet, Sprint
.fn3	Font, Harvard Graphics 3.0
.fnt	Font
.fnx	Inactive Font, Exact
.fo1	Font, Borland Turbo C
.fo2	Font, Borland Turbo C
.foc	Focus
.fon	Font
.for	Fortran Program File, MS Fortran
.fot	Truetype Font, MS Windows
.fp	Configuration, Foxpro
.fpc	Catalog, Foxpro
.fpq	Foxpro Queries, Foxpro
.fpt	Memo, Foxpro
.fpw	Configuration, Foxpro
.fr3	Dbase IIi+ Form File, Dbase IV
.frg	Uncompiled Report File, Dbase IV
.frm	Form, Visual Basic
.frm	Report File, Clipper 5/dBASE IV/dBFAST
.frm	Text Form
.fro	Compiled Report File, Dbase IV
.frs	Screen Characters, Wordperfect For Windows
.frt	Report Memo, Foxpro
.frx	Graphic In Vb-Form, Visual Basic
.frx	Report, Foxpro
.fsl	Dynamic Link Library
.fssd	Sound File
.fst	Linkable Program, dBFAST
.fsx	Lotus 1-2-3
.ft	File Transfer Configuration, IRMA Workstation For Windows
.ftm	Font, MicroGrafx
.ftp	Ftp-Configuration
.ful	Complete Back-Up
.fw	Database, Framework
.fw2	Database, Framework II
.fw3	Database, Framework III
.fxd	Foxdoc
.fxp	Ordered File, Foxpro
.fxs	Graphic, WINFax
.gb	Chinese
.gdf	Dictionary File, Geos
.gds	Image File
.gem	Vector Graphic, GEM/Ventura

	Publisher
.gen	Generated Text, Ventura Publisher
.gen	Ordered Template, dBASE
.geo	Geoworks
.gfb	Compressed Gif-Bild, Gifblast
.gif	Picture File
.gif	Bitmap Graphic, Graphics Interchange Format
.gl	Animated Picture
.gl	Animation, GRASP
.gly	Glossary, MS Word
.gmf	Cgm Graphic, Applause
.gr2	Screen Driver, MS Windows
.gr3	Windows 3.0 Screen Grabber
.gra	MS Flight Simulator
.grb	Ms-DOS Shell Monitor
.grf	Graph, Charisma/GRAPH Plus
.grp	Program Manager Group, MS Windows
.gry	Picture File
.gz	Compressed
.gz	Compressed Archive File, GZIP
.gzip	Compressed
.h	Text
.h	Header File
.ha	Compressed
.ha	Compressed Archive File, HA
.ham	Image File
.hbk	Handbook, MathCad
.hdl	Alternate Download File Listing, Procomm Plus
.hdr	Message Header, Procomm Plus
.hdr	PC-File+ Database Header
.hdx	Help Index
.hex	Hex Dump
.hfi	Hp Font Info, GEM
.hgl	Graphic, HP Graphics Language
.hh	C++ Header
.hhh	Precompiled Header, Power C
.hhp	Help Data, Procomm Plus
.hlb	VMS Help Libraries
.hlp	VMS Help Files
.hlp	Help Data
.hmm	Alternate Mail Read Option Menu, Procomm Plus
.hof	Hall of Fame, distinct scores
.hpf	Hp Laserjet Fonts, Pagemaker
.hpg	Vector Graphic, HPGL Plotter file
.hpi	Fonts Information, GEM
.hpj	Help Project, MS Help Compiler
.hpk	Compressed
.hpk	Compressed Archive File, HPack
.hpm	Alternate Main Menu For Privileged
	Users, Procomm Plus
.hpp	C++ Header, Zortech C++
.hqx	Macintosh Compressed
.hqx	Compressed Archive File Macintosh ASCII, BinHex
.hrm	Alternate Main Menu For Limited/Normal Users, Procomm plus
.hrz	Image File
.hst	History, Procomm Plus
.htm	Hypertext Markup Language, HTML
.htx	Hypertext File
.hxm	Alternate Protocol Selection Menu For All Users, Procomm plus
.hxx	C++ Header
.hyc	Wordperfect
.hyd	Hyphenation Dictionary File, Wordperfect For Windows
.hyp	Compressed
.hyp	Compressed Archive File, Hyper
.hz	Chinese
.ibm	Picture File
.ice	Archive
.icl	Icon Library, PC Tools
.icn	Icon Program File
.ico	Image File
.ico	Icon, MS Windows
.id	Disk Identification File
.idf	Instruments Drivers File
.idx	Index
.ief	Image File
.iff	Several Types of Binary Data
.iff	Graphic, Sound, Interchangeable File Format
.ifs	System File, OS/2
.igs	Iges-Format
.im8	Sun Raster Graphic
.img	Picture File
.img	Bitmap Graphic, GEM/Ventura Publisher
.in3	Input Device Driver, Harvard Graphics 3.0
.inc	Include File, diverse Programmer's Language
.ind	Index, Dbase IV
.inf	Help File, OS/2 Buch
.inf	Information Text File, ASCII
.inf	Setup File
.ini	Initialization File
.ink	Pantone Reference File, Corel Draw
.ins	Wordperfect
.int	Borland Interface Units
.int	Code Page File, collation sequence,

	Foxpro
.inx	Index, Foxbase
.io	Compressed Archive File, CPIO
.ion	Text Program/Installation Description File, 4DOS & others
.ipl	Pantone Spot Reference Palette, Corel Draw
.irs	Wordperfect
.isd	Dictionary File, RapidFile
.ish	Compressed
.ish	Compressed Archive File, ISH
.iw	Iconware Application, Icon Author
.iwa	Text File, IBM Writing Assistant
.iwp	Text File, Wang
.jas	Image File
.jas	Graphic
.jbig	Image File
.jfi	Image File
.jis	Japanese
.job	Job File, BETA 44
.jor	Journal File Sql
.jpc	Picture File
.jpg	Picture File
.jpg	Graphic Jpeg, Joint Photography Experts Group
.jw	Text File, JUSTWrite
.jwl	Text File Library, JUSTWrite
.jzz	Spreadsheet File, Jazz
.kbd	Keyboard Layout, IRMA Workstation For Windows
.kbm	Keyboard Layout
.kex	Kedit Profile File
.key	Keyboard Macros
.kyb	Keyboard Layout
.l	Lex Program File
.l	Link Instructions
.l	Lisp Program File
.lab	Mailing Labels, Q+E For MS Excel
.lay	Word Chart Layout, Applause
.lbg	Label Generator, Dbase IV
.lbl	Label, dBASE IV/dBFAST/Clipper 5
.lbm	Image File
.lbm	Bitmap Graphic, Deluxe Paint
.lbo	Ordered Labels, Dbase IV
.lbr	Archive
.lbr	Compressed Archive File, LU
.lbr	Screen Driver, Lotus 1-2-3
.lbt	Label Memo, Foxpro
.lbx	Label, Foxpro
.lcf	Linker Control File, Norton GuideS
.lcn	Dictionary File, Wordperfect
.ld	Long Distance Codes, Telix
.ld1	Overlay File, dBASE

.ldb	Database File, MS Access
.les	Learn Program, LESson
.lex	Lexicon, Dictionary file
.lft	Laserprinter Font, CHIWriteR
.lgo	MS Windows Logo Driver, MS Windows
.lha	Compressed
.lhw	Compressed Archive File Amiga, Lhwarp
.lib	Program Library, diverse Programmer's Language
.lif	Compressed Archive File
.lif	Logical Interchange Format Data File, Hewlett-Packard
.lin	Line Art, AutoCAD
.lis	Program Listing
.lj	Text File For Hp JL II Printer
.lm8	Picture File
.lnk	Linker File
.log	Log File
.lrf	Linker Response File, MS C/C++
.lrs	Language Resource File, Wordperfect For Windows
.lsp	Lisp Program File, XLISP
.lst	List File
.lst	List of Files On Installation Disks, MS-DOS and others
.lzh	Compressed
.lzh	Compressed Archive File, LHA/LHARC
.lzs	Compressed
.lzs	Compressed Archive File, LARC
.lzw	Lhwarp Archive
.lzw	Compressed Archive File Amiga, Lhwarp
.lzx	Compressed
.m	Mathematica
.m	Function, Matlab
.m	Macro Sourcecode, Brief
.m3	Modulea 3 Program File
.m3d	3d Animationsmacro
.ma3	Macro, Harvard Graphics 3.0
.mac	Bitmap Graphic, MACPaint
.mac	Macro
.mac	Picture File
.mag	Picture File
.mak	Make File
.mak	Project File, Visual Basic
.man	Manual
.man	Text Manual
.map	Image File
.map	Colorpalette
.map	Format Data, Micrografx Picture

	Publisher
.map	Linker Map File
.mas	Master Data Definition, Focus
.mat	Sound File
.mat	Matlab
.max	Max Program File
.mb	Memo Holder For Database, Paradox
.mbk	Multiple Index Back-Up, Dbase IV
.mcc	Configuration, MathCad
.mcd	Document, MathCad
.mcf	Font, MathCad
.mci	Mci Command Set
.mcp	Printer Driver, MathCad
.mcw	Text File, Word For MacIntosh
.md	Compressed
.md	Compressed Archive File, MDCD
.mdb	Database, MS Access
.mdm	Modem Definition, Telix
.mdx	Multiple Index, Dbase IV
.me	Standard: Read.Me - ASCII Text File
.meb	Macro Editor Bottom Overflow File, Wordperfect
.med	Macro Editor Delete Save, Wordperfect
.mem	Macro Editor Macro, Wordperfect
.mem	Memory File of Variables, Memory, dBASE IV/Foxpro
.meq	Macro Editor Print Queue, Wordperfect
.mer	Macro Editor Resident Area, Wordperfect
.mes	Macro Editor Work Space, Wordperfect
.met	Macro Editor Top Overflow File, Wordperfect
.met	Meta File
.meu	Menus, DOS Shell
.mex	Mex File, executable command, Matlab
.mf	Metafont Text File
.mgf	Image File
.mgf	Font, MicroGrafx
.mhg	Multimedia
.mid	Music File
.mid	Midi- File, MS Windows
.mif	Midi Instrument
.mix	Object File, Power C
.mk	Makefile
.mke	Makefile, MS Windows SDK
.mki	Picture File
.mlb	Macro Library, Symphony, Foxpro
.mmm	Animation
.mnd	Menu Program File, AutoCAD
.mnt	Menu Memo, Foxpro
.mnu	Menu, AutoCAD/Norton Commander
.mnx	Compiled Menu File, AutoCAD
.mnx	Menu, Foxpro
.mob	Device Definitions, Pen Windows
.mod	Music File
.mod	Audio File, Amiga
.mod	Modulea-2 Program File
.mod	Windows Kernel Module
.mov	Movie
.mp	Multiplan
.mp2	Audio
.mpa	Audio
.mpc	Calender, MS Project
.mpd	Mini Port Driver
.mpg	Multimedia
.mpg	Mpeg Animation
.mpm	Mathplan Macro, Wordperfect
.mpp	Project File, MS Project
.mpr	Program, Foxpro
.mps	Multimedi
.mpv	View, MS Project
.mpx	Compiler Menu, Foxpro
.mrb	Image File
.mrb	Multiple Resolution Bitmap Graphic, MS C/C++
.mrs	Macro Resource File, Wordperfect For Windows
.ms	Checksum File For Anti-Virus, MS-DOS
.msg	Message, Message Text file, ASCII
.msp	Image File
.msp	Bitmap Graphic, MS Windows 2.x Paint
.mst	Minispecification, Prosa
.mst	Setup Script, MS Windows SDK
.msw	Text File, MS Word
.mth	Math, Derive
.mtm	Sound File
.mtv	Picture File
.mtw	Data File, Minitab
.mu	Menu, Quattro Pro
.mus	Audio File
.mvb	Database, MS Info Viewer
.mvf	Stop Frame File, AutoCAD Autoflix
.mvi	Movie Command File, AutoCAD Autoflix
.mxt	MS C
.nb	Text File, Nota Bene
.ncd	Norton Change Directory, Norton Commander
.ndx	Index, dBASE/dBFAST
.net	Network Configuration, Harvard Graphics For Windows/Paradox
.new	New Info

.nff	Image File
.ng	Online-Documentation, Norton Guide
.nls	National Language Support - Code Page
.not	Notes, IBM Bookmanager
.nst	Amima Mod Audio File
.nst	Sound File
.ntf	Notes, IBM Bookmanager
.nts	Tutorial, Norton
.ntx	Index, Clipper 5
.nws	Info Text File, Latest News, ASCII
.o	Object File, UNIX
.o$$	Outfile, Sprint
.obj	Object Code
.ocx	Ole Control Extension
.off	Image File
.off	Vector Graphic, Object File Format
.ofm	Font, Adobe
.olb	Object Library
.old	Back-Up
.oli	Text File, Olivetti
.omf	Image File
.opc	Job File, BETA 44
.opt	Optimize, QEMM
.opx	Organization Diagram, MS ORGChart
.out	Output File
.ov1	Program File - Overlay
.ov2	Program File - Overlay
.ov3	Program File - Overlay
.ovl	Program File - Overlay
.ovr	Program File - Overlay
.p	Pascal Program File
.p	Picture, Applause
.pad	Keypad Definition, TELEMATE
.pak	Compressed Archive File, PAK
.pal	Color Palette
.par	Parameter File, Fractint
.pas	Pascal Program File
.pat	Hatch-Pattern, AutoCAD
.pat	Vector-Pattern, Corel Draw
.pbi	Profiler Binary Input, MS Source Profiler
.pbm	Picture File
.pbm	Pbm Portable Bit Map Graphic
.pbo	Profiler Binary Output, MS Source Prof
.pbt	Profiler Binary Table, MS Source Profiler
.pc	Text File Ibm PC-Specific Info
.pc3	Custom Palette, Harvard Graphics 3.0
.pc8	ASCII Text Ibm-8
.pcc	Image File
.pcd	Kodak Photo Cd
.pcd	Graphic, Photo-CD Image

.pcf	Profiler Command File, MS Source Profiler
.pch	Patch File
.pch	Precompiled Header, MS C/C++
.pck	Pick File, Turbo Pascal
.pcl	Hp-PCl Graphic Data File, HP Printer Control Language
.pcm	Sound File
.pct	Picture File
.pct	Bitmap Graphic
.pcw	Text File, PC Write
.pcx	Picture File(Paintbrush
.pcx	Bitmap Graphic, Paintbrush
.pdb	Tact
.pdf	Image File
.pdi	MS Powerpoint
.pdr	Port Driver
.pdr	Printer Driver
.pds	Pds Graphic
.pdv	Printer Driver, Paintbrush
.peb	Program Editor Bottom Overflow File, Wordperfect
.ped	Program Editor Delete Save, Wordperfect
.pem	Program Editor Macro, Wordperfect
.peq	Program Editor Print Queue File, Wordperfect
.per	Program Editor Resident Area, Wordperfect
.pes	Program Editor Work Space File, Wordperfect
.pet	Program Editor Top Overflow File, Wordperfect
.pf	Profile File, Monitor, Printer
.pfa	Outline Font
.pfb	Outline Font
.pfk	Programmable Function Keys, Xtree Pro
.pfm	Printer Outline Font Metric: Font, Adobe Type 1 PostScript
.pft	Printer Font, CHIWriteR
.pgm	Pbm Portable Gray Map Graphic
.pgp	Pretty Good Privacy
.ph	Perl Header
.ph	Phrase-Table, MS C/C++, MS Help COMpiler
.pic	Bitmap Graphic
.pict	Picture File, Macintosh
.pif	Program Information File, MS Windows
.pit	Macintosh Compressed
.pit	Compressed Archive File Macintosh, PACKIT

.pjt	Project Memo, Foxpro
.pjx	Project, Foxpro
.pka	Compressed Archive File, PKARC
.pl	Palette, Harvard Graphics For Windows
.pl	Perl Program File
.pl3	Chart Palette, Harvard Graphics 3.0
.plb	Library, Foxpro
.plc	Add-In, Lotus 1-2-3
.pln	Spreadsheet File, Wordperfect For Windows
.plt	HPgl Plotter File Vector Graphic, AutoCAD
.pm	Picture File
.pm	PM Graphic
.pm3	Document, Pagemaker 3
.pm4	Document, Pagemaker 4
.pm5	Document, Pagemaker 5
.pn3	Printer Driver, Harvard Graphics 3.0
.png	Image File
.pnm	Pbm Portable Any Map Graphic
.pnt	Macpaint
.pnt	Macintosh Painting
.pol	Music File
.pop	Popup Menu, dBASE
.pp	Amiga Compressed
.pp	Compressed Archive File Amiga, Powerpacker
.ppd	Postscript Printer Description
.ppl	Polaroid Palette Plus Colorkey Device Driver, Harvard graphics 3.0
.ppm	Video
.ppm	Pbm Portable Pixel Map Graphic
.ppo	Preprocessor Output, Clipper 5
.ppt	Powerpoint-Foils, Powerpoint
.pr2	Presentation, Aldus Persuasion 2.0
.pr2	Printer Driver, Dbase IV
.pr3	Postscript Printer Driver, Dbase IV
.prd	Printer Driver, Diverse
.pre	Settings, preferences, MS C/C++
.prf	Printer Driver, Dbase IV
.prf	Profiler Output
.prg	Program File, Clipper 5/dBASE IV/dBFAST/Foxpro
.prj	Project
.prm	Parameter, Diverse
.prn	Text File, Lotus 1-2-3/Symphony
.pro	Prolog Program File
.prs	Printer Resource File For Fonts, Wordperfect For Windows
.prs	Procedure, Dbase IV
.prs	Harvard Graphics
.prt	Printer Driver, DR. Halo
.prt	Printer Configuration, IRMA Workstation For Windows
.prx	Compiler Program, Foxpro
.ps	Laser Printer File
.ps	Postscript, ASCII
.ps/.psid	Picture File/Text
.psf	Postscript Printer Font, CHIWriteR
.psm	Symbol Table, Turbo Pascal
.pt3	Device Driver, Harvard Graphics 3.0
.pt3	Template, Pagemaker 3
.pt4	Template, Pagemaker 4
.pt5	Template, Pagemaker 5
.pub	Public Key Ring, Pretty Good Privacy
.pub	Publication, Ventura Publisher
.pw	Text File, Professional Write
.pwl	Password List
.px	Primer Database Index, Paradox
.py	Python Script
.pyc	Compiler Python Script
.pzd	Pizazz Plus
.pzo	Pizazz Plus
.qag	Quick Access Group, Norton Desktop
.qbd	Keyboard Layout, IRMA Workstation For Windows
.qbe	Example Query, Query By Example, dBASE IV/Quattro pro
.qbo	Ordered Query, Dbase IV
.qef	Query, Q+E For MS Excel
.qep	Irma Workstation For Windows
.qlb	Quick Library, MS C/C++
.qlc	Postscript Help File
.qpr	Driver For Printer Queue, OS/2
.qpr	Relational Query, Foxpro
.qpx	Ordered Query, Foxpro
.qrs	Equation Editor, Wordperfect For Windows
.qrt	Qrt Ray Tracing Graphic
.qry	Query, query, Dbase IV
.r8	Raw Graphic
.ras	Picture File
.ras	Sun Raster Graphic
.raw	Animated Picture
.raw	Raw Rgb 24-Bit Graphic
.rax	Image File
.rbf	Data File, RBASE
.rc	Resource Compiler Script File, Borland C++/MS C++
.rdx	Data File, Reflex
.rec	Record File, Sprint
.rec	Recorded Macro, MS Windows
.ref	Cross Reference
.reg	Registration Data, MS

	Windows/Regedit
.rem	Annotation
.req	Request
.res	Compiler Resources, Borland C++/MS C++
.res	Dbase Resources, Dbase IV
.rex	Rexx Program File
.rez	Resources
.rf	Sun Raster Graphic
.rft	Text File, DCA/RFT Revisable Format Text, IBM displaywrite
.rgb	Picture File
.ri	Lotus 1-2-3
.rlb	Harvard Graphics For Windows
.rle	Picture File
.rle	Graphic, Run Length Encoded
.rmi	Sound File
.rmi	Midi File
.rmk	Makefile, RMAKE
.rnd	Rendering Slide, AutoCAD Autoshade
.rol	Sound File
.rol	Adlib Music, Roland
.rpd	Database, Rapidfile
.rpt	Report
.rr	Railroad Tycoon
.rsl	Resources Library, PC Tools
.rsp	Response, Text file and Eingaben
.rtf	Rich Text Format Text File
.rtf	Windows Help File Script
.rtl	Text File
.run	Script Tools/Program, PC Tools
.s$$	Temporary Sort File, Sprint
.s3m	Sound File
.sam	Text File, Samna, AMI Pro 2.x, 3.x
.sav	Back-Up
.sb	Audio File
.sbp	Dml Program, SuperBASE 4
.sc	Screen Driver, Framework II
.sc3	Screen Driver, Harvard Graphics 3.0
.scc	Text File
.scd	Graphic, SCODL Scan Conversion Object Description language
.sch	MS Schedule
.scn	Compressed Screen Format
.sco	High Score, diverse Programs
.scr	Fax, Faxview
.scr	Screen Saver, MS Windows
.scr	Screen Snapshot, diverse Programs
.scr	Script
.sct	Screen, Foxpro
.sct	Script Tools - Program File, PC Tools
.scx	Image File File
.scx	Screen, Foxpro

.sdf	System File Format
.sdn	Shell Archive
.sea	Self-Extracting, Mac
.sea	Compressed Archive File Macintosh
.sec	Compressed Archive File
.sec	Secret Key Ring, Pretty Good Privacy
.seq	Animation, Atari
.set	Configuration, diverse Back-ups
.sf	Sound File
.sf	Wps Attribute Storage, OS/2
.sfi	Printer Font, HP LaserJet landscape, Ventura Publisher
.sfl	Printer Font, HP LaserJet landscape, Ventura Publisher
.sfp	Printer Font, HP LaserJet portrait, Ventura Publisher
.sft	Screenfont, CHIWriteR
.sgi	Picture File
.sh	Archive File, Shar
.sh3	Presentation, Harvard Graphics 3.0
.shar	Shell Archive
.shb	Background File, Corel Show
.shg	Image File File
.shg	Hypergraphic, Hotspot Editor
.shk	Compressed, Mac
.shk	Compressed Archive File Apple II, Shrinkit
.shm	Shell Macro, Wordperfect
.shr	Archive File, SHAR
.shw	Presentation/Show File, Harvard Graphics 2.0/Corel Show
.sik	Back-Up File
.sim	Simulation, MS Flight Simulator
.sit	Compressed, Mac
.sit	Compressed Archive File Macintosh, Stuffit
.sl	S-Lang Program File
.slb	Slide Library, AutoCAD
.slc	Compiler Salt Script, Telix
.sld	Slide, AutoCAD
.slk	Sylk, Symbolic Link, Multiplan
.slt	Salt Script Application Language For Telix Script Source, Telix
.sm	Smalltalk Program File
.sm	Text File, Samna Word
.smm	Macro, AMI Pro 2.x, 3.x
.smp	Sample
.snd	Music File File
.snd	Sound File
.sng	Song, MIDI
.sno	Snobol Program File
.sp	Compressed Archive File, Splint
.spc	Program, MS Multiplan

.spc	Temporary File, Wordperfect For Windows
.spd	Bitstream Typefaces Font, Harvard Graphics 3.0
.spd	Postscript Printer Mini Driver
.spg	Glossary, Sprint
.spl	Compressed Archive File, Splint
.spl	Printer Driver, Sprint
.spl	Printer -Spool File, MS Windows
.spl	Sample
.spm	Wordperfect
.spp	Printer File, Sprint
.spr	Document Letter, Sprint
.spr	Generated Screen Program, Foxpro
.sps	Screen Driver, Sprint
.spt	Spitbol Program File
.spx	Compiler Screen Program, Foxpro
.sql	Sql Data
.sqz	Squeeze For Msdos
.sqz	Compressed Archive File, Squeeze
.src	Sourcecode, Source, Dataflex and others
.srf	Graphic, Sun Raster File
.st	Smalltalk Program File
.std	State Transition Diagram Graphic, Prosa
.stf	Compressed Archive File, Shrinktofit
.stf	Setup Information
.stm	Sound
.stm	State Transition Diagram Model, Prosa
.str	Structure List, dBASE
.sts	Project Status Info, MS C/C++
.sty	Template, Style library/sheet
.sun	Graphic, Sun rasterfile
.sup	Supplementary Dictionary File, Wordperfect For Windows
.svd	Back-Up File, MS Word
.svg	Back-Up File, MS Word
.svs	Back-Up File, MS Word
.sw	Audio File, signed word
.swp	Back-Up File, Sprint
.swp	Swap File, DOS
.sy3	Symbol File, Harvard Graphics 3.0
.syd	Back-Up File, MS Windows Sysedit.exe
.sym	Symbol File, Harvard Graphics/MS Windows SDK
.sym	Symbol Table, diverse Compiler
.syn	Synonyms, MS Word 5
.sys	System File: Device Driver, HardwareConfiguration
.syw	Graphic, Symbol, Harvard Graphics For Windows
.tag	Query Tag Name, Dataflex
.tah	Help File For Turbo Assembler, Borland C++
.tar	Tape Archive
.tar	Compressed Archive File, TAR
tar.gz	Compressed TAR
tar.z	Compressed TAR
.tb1	Font, Borland Turbo C
.tb2	Font, Borland Turbo C
.tbk	Memo Back-Up, dBASE IV/Foxpro
.tbk	Toolbook, Asymetrix ToolBook
.tc	Configurations File, Borland C++/Turbo C
.tch	Help File, Borland C++/Turbo C
.td	Configurations File, Turbo Debugger
.tdb	Database
.tdf	Outline Font
.tdh	Help File Turbo Debugger, Borland C++
.tdk	Keyboard Record, Turbo Debugger
.tdo	Compressed
.tds	Symbol Table, Turbo Debugger
.tem	Template, Icon Author
.tem	Turbo Editor Macro Language, Borland C++
.tex	Tex Text File
.tf	Configurations File, Turbo Profiler
.tfa	Area File, Turbo Profiler
.tfh	Help File Turbo Profiler, Borland C++
.tfs	Statistik, Turbo Profiler
.tga	Picture
.tga	Truevision Targa Graphic
.thb	Graphic, Bitmap
.thm	Database Der Graphicen, MS Clipart Gallery
.ths	Thesaurus Dictionary File, Wordperfect For Windows
.tif	Tagged Image File Format, Bitmap Graphic
.tif(f	Image File File
.tmf	Tagged Font Metric File, Wordperfect For Windows
.tmo	ZTG Global Optimizer Default Output File, Zortech C++
.tmp	Temporary File
.tms	Script, Telemate
.toc	Table of Contents Text File, ASCII
.tp	Configurations File, Turbo Pascal
.tp3	Template, Harvard Graphics 3.0
.tph	Help File, Turbo Pascal
.tpl	Resident Units, Turbo Pascal

.tpl	Template, Harvard Graphics 2.0	.voc	Sound File
.tpp	Protected Mode Units, Turbo Pascal	.voc	Audio File
.tpu	Turbo Pascal Unit, BGI, Turbo Pascal	.vsd	Drawing, VISIO
.tpw	Turbo Pascal Unit, BGI, Turbo Pascal For Windows	.vss	Stencil, VISIO
.tre	Directory Tree, PC Tools	.vst	Template, VISIO
.trm	Terminal Settings, MS Windows	.vst	Truevision Vista Graphic
.trn	Translation, Quattro	.vue	View, dBASE IV/Foxpro
.tsp	Windows Telephony Service Provider	.vw	Text File, VOLKSWriteR
.tst	Printer -Test File, Wordperfect For Windows	.vwr	File Viewer File, PC Tools
.ttf	Truetype Font	.vxd	Virtual Device Driver
.tut	Tutorial	.w	Word Chart, Applause
.tx8	DOS-Text	.w30	Printer Font, AST TurboLaser, Ventura Publisher
.txt	Text File	.w40	Back-Up File From MS Windows 95
.tym	Time Stamp Program, Pagemaker v4 & v5	.wav	Waveform Audio File, sound
.ub	Audio File, unsigned byte	.wbf	Windows Batch File
.uc2	Compressed	.wbk	Workbook, Wordperfect For Windows
.ui	User Interface, Sprint	.wcd	Macro Token List, Wordperfect For Windows
.ul	Sound	.wcm	File Transmission File, MS Works
.ul	Ulaw Audio File	.wcm	Macro, Wordperfect For Windows
.uld	Uploaded Files, Procomm Plus	.wdb	Database, MS Works
.unx	Text File - UNIX Specific Info	.wfn	Graphics: Symbols And Fonts, Corel Draw
.upd	Update File, dBASE	.wfx	Data File, WINFax
.upo	Compiled Update File, dBASE	.wid	Width Table, Ventura Publisher
.usp	Printer Font Us-ASCII Enhanced Characters, Pagemaker	.win	Window, dBASE/Foxpro
.usr	User Database, Procomm Plus/Turbo C++	.wk1	Spreadsheet File, Lotus 1-2-3 2.x/Symphony
.utl	Sound File	.wk3	Spreadsheet File, Lotus 1-2-3 3.x
.uu	Compressed Archive File ASCII, UUENCODE	.wkb	Document, Wordperfect For Windows
.uud	Uudecoded	.wke	Spreadsheet File, Lotus 1-2-3, Student version
.uue	Uuencoded	.wkq	Spreadsheet File, Quattro Pro
.uue	Compressed Archive File ASCII, UUENCODE	.wks	Spreadsheet File, Lotus 1-2-3/Symphony/MS Works
.uw	Audio File, unsigned word	.wks	Workspace, XLISP
.val	Value, values, dBASE	.wmc	Macro, Wordperfect For Windows
.var	Variable, Icon Author	.wmc	Text File, WordMARC
.vbx	Visual Basic Extension, Visual Basic	.wmf	Vector Graphic, Windows MetaFile, MS Windows
.vc	Spreadsheet File, Visicalc	.wn	Text File, Write Now
.vcx	Spreadsheet File, Visicalc	.wnf	Outline Font, Corel Draw
.vew	Viewfile, Clipper 5	.woa	Swap File, MS Windows
.vga	Vga Screen Characters, Diverse	.wp	Text File, Wordperfect v4.2
.vga	Vga Screen Driver, Diverse	.wp5	Text File, Wordperfect v5
.vgr	Graphic, Ventura Publisher	.wpd	Screen Driver, Wordperfect
.vic	Picture File	.wpf	Text File, Wordperfect
.vid	Monitor Control File, MS-DOS Shell	.wpg	Wordperfect Graphic Bitmap, DrawPerfect
.vid	Screen Driver, Word	.wpk	Macros, Wordperfect For Windows
.vik	Picture File	.wpm	Macros, Wordperfect
.vis	Picture File	.wps	Text Document, MS Works
.vmf	Font Characteristics, Ventura Publisher		

.wq!	Compressed Spreadsheet File, Quattro Pro
.wq1	Spreadsheet File, Quattro Pro
.wr1	Spreadsheet File, Symphony
.wrd	Template, CHARISMA
.wri	Text File, Windows Write
.wrk	Spreadsheet File, Symphony 1.0
.wrp	Compressed Archive File Amiga, Warp
.wrs	Windows Printer Driver, Wordperfect For Windows
.ws	Text File, WordSTAR 5.0+6.0
.ws2	Text File, WordSTAR 2000
.wsd	Data File, WordSTAR
.wst	Text File, WordSTAR
.x	Lex Program File
.x01	Secondary Index, Paradox
.x02	Secondary Index, Paradox
.x03	Secondary Index, Paradox
.x04	Secondary Index, Paradox
.x05	Secondary Index, Paradox
.x06	Secondary Index, Paradox
.x07	Secondary Index, Paradox
.x08	Secondary Index, Paradox
.x09	Secondary Index, Paradox
.xab	Address Book, XENIX MAIL
.xbm	Picture, X Windows
.xbm	Bitmap Graphic
.xfn	Printer Font, Ventura Publisher
.xft	24 Pin Printer Font, CHIWriteR
.xla	Addin: Macro, MS Excel
.xlb	MS Excel
.xlc	Chart Diagram, MS Excel
.xll	Excel Dynamic Link Library, MS Excel
.xlm	Macrotemplate, MS Excel
.xls	Spreadsheet File, MS Excel
.xlt	Mastertemplate, MS Excel
.xlt	Translation Table, Lotus 1-2-3/Symphony
.xlw	Workspace, MS Excel
.xm	Sound
.xpm	Picture File
.xpm	Graphic
.xqt	Macro Sheet, SuperCALC
.xrf	Cross Reference
.xwd	Image File
.xy	Text File, XY Write
.xy3	Text File, XYWrite III
.xyw	Text File, XYWrite III
.y	Compressed Archive File Amiga, YABBA
.y01	Secondary Index, Paradox
.y02	Secondary Index, Paradox
.y03	Secondary Index, Paradox
.y04	Secondary Index, Paradox
.y05	Secondary Index, Paradox
.y06	Secondary Index, Paradox
.y07	Secondary Index, Paradox
.y08	Secondary Index, Paradox
.y09	Secondary Index, Paradox
.yuv	Image File
.z	Compressed Archive File ASCII, Compress
.zip	Compressed
.zip	Compressed Archive File, PKZIP
.zom	Compressed Archive File Amiga, Zoom
.zoo	Compressed Archive File, ZOO
.zvd	Zyxel Voice File, Z-Fax
.zw	Chinese

Appendix E — TCP/UDP System Port Assignments

Port	TCP	UDP	Function	Description	Comment
0	tcp			Reserved	
0		udp		Reserved	
1	tcp		tcpmux	TCP Port Service Multiplexer	rcpmux; rarely used
1		udp	tcpmux	TCP Port Service Multiplexer	rcpmux; rarely used
2	tcp		compressnet	Management Utility	
2		udp	compressnet	Management Utility	
3	tcp		compressnet	Compression Process	
3		udp	compressnet	Compression Process	
4	tcp			Unassigned	
4		udp		Unassigned	
5	tcp		rje	Remote Job Entry	
5		udp	rje	Remote Job Entry	
6	tcp			Unassigned	
6		udp		Unassigned	
7	tcp		echo	Echo	echo; see if a host is reachable (ping)
7		udp	echo	Echo	echo; see if a host is reachable (ping)
8	tcp			Unassigned	
8		udp		Unassigned	
9	tcp		discard	Discard	discard
9		udp	discard	Discard	discard
10	tcp			Unassigned	
10		udp		Unassigned	
11	tcp		systat	Active Users	systat
11		udp	systat	Active Users	systat
12	tcp			Unassigned	
12		udp		Unassigned	
13	tcp		daytime	Daytime	daytime
13		udp	daytime	Daytime	daytime
14	tcp			Unassigned	
14		udp		Unassigned	
15	tcp			Unassigned [was netstat]	netstat
15		udp		Unassigned	netstat
16	tcp			Unassigned	
16		udp		Unassigned	
17	tcp		qotd	Quote of the Day	qotd; quote of the day
17		udp	qotd	Quote of the Day	qotd; quote of the day
18	tcp		msp	Message Send Protocol	
18		udp	msp	Message Send Protocol	
19	tcp		chargen	Character Generator	chargen; character generator
19		udp	chargen	Character Generator	chargen; character generator
20	tcp		ftp-data	File Transfer [Default Data]	
20		udp	ftp-data	File Transfer [Default Data]	
21	tcp		ftp	File Transfer [Control]	ftp (file transfer protocol); copies files, renames, deletes

21		udp	ftp	File Transfer [Control]	ftp (file transfer protocol); copies files, renames, deletes
22	tcp			Unassigned	
22		udp		Unassigned	
23	tcp		telnet	Telnet	telnet; allows clients and servers to speak, including access to router
23		udp	telnet	Telnet	telnet; allows clients and servers to speak, including access to router
24	tcp			any private mail system	private mail
24		udp		any private mail system	private mail
25	tcp		smtp	Simple Mail Transfer	smtp (simple mail transfer protocol); restrict to single e-mail server
25		udp	smtp	Simple Mail Transfer	smtp (simple mail transfer protocol); restrict to single e-mail server
26	tcp			Unassigned	
26		udp		Unassigned	
27	tcp		nsw-fe	NSW User System FE	
27		udp	nsw-fe	NSW User System FE	
28	tcp			Unassigned	
28		udp		Unassigned	
29	tcp		msg-icp	MSG ICP	
29		udp	msg-icp	MSG ICP	
30	tcp			Unassigned	
30		udp		Unassigned	
31	tcp		msg-auth	MSG Authentication	
31		udp	msg-auth	MSG Authentication	
32	tcp			Unassigned	
32		udp		Unassigned	
33	tcp		dsp	Display Support Protocol	
33		udp	dsp	Display Support Protocol	
34	tcp			Unassigned	
34		udp		Unassigned	
35	tcp			any private printer server	
35		udp		any private printer server	
36	tcp			Unassigned	
36		udp		Unassigned	
37	tcp		time	Time	time
37		udp	time	Time	time
38	tcp			Unassigned	rap; route access protocol
38		udp		Unassigned	rap; route access protocol
39	tcp		rlp	Resource Location Protocol	
39		udp	rlp	Resource Location Protocol	
40	tcp			Unassigned	
40		udp		Unassigned	
41	tcp		graphics	Graphics	
41		udp	graphics	Graphics	

42	tcp		nameserver	Host Name Server	nameserver
42		udp	nameserver	Host Name Server	nameserver
43	tcp		nicname	Who Is	nicname; whois; outbound only or block
43		udp	nicname	Who Is	nicname; whois; outbound only or block
44	tcp		mpm-flags	MPM FLAGS Protocol	
44		udp	mpm-flags	MPM FLAGS Protocol	
45	tcp		mpm	Message Processing Module [recv]	
45		udp	mpm	Message Processing Module [recv]	
46	tcp		mpm-snd	MPM [default send]	
46		udp	mpm-snd	MPM [default send]	
47	tcp		ni-ftp	NI FTP	
47		udp	ni-ftp	NI FTP	
48	tcp			Unassigned	auditd
48		udp		Unassigned	auditd
49	tcp		login	Login Host Protocol	login; tacacs-65
49		udp	login	Login Host Protocol	login; tacacs-65
50	tcp		re-mail-ck	Remote Mail Checking Protocol	
50		udp	re-mail-ck	Remote Mail Checking Protocol	
51	tcp		la-maint	IMP Logical Address Maintenance	
51		udp	la-maint	IMP Logical Address Maintenance	
52	tcp		xns-time	XNS Time Protocol	
52		udp	xns-time	XNS Time Protocol	
53	tcp		domain	Domain Name Server	dns (domain name service); contains host names; can be spoofed
53		udp	domain	Domain Name Server	dns (domain name service); contains host names; can be spoofed
54	tcp		xns-ch	XNS Clearinghouse	
54		udp	xns-ch	XNS Clearinghouse	
55	tcp		isi-gl	ISI Graphics Language	
55		udp	isi-gl	ISI Graphics Language	
56	tcp		xns-auth	XNS Authentication	
56		udp	xns-auth	XNS Authentication	
57	tcp			any private terminal access	
57		udp		any private terminal access	
58	tcp		xns-mail	XNS Mail	
58		udp	xns-mail	XNS Mail	
59	tcp			any private file service	
59		udp		any private file service	
60	tcp			Unassigned	
60		udp		Unassigned	
61	tcp		ni-mail	NI MAIL	
61		udp	ni-mail	NI MAIL	
62	tcp		acas	ACA Services	
62		udp	acas	ACA Services	
63	tcp		via-ftp	VIA Systems - FTP	
63		udp	via-ftp	VIA Systems - FTP	
64	tcp		covia	Communications Integrator (CI)	

64		udp	covia	Communications Integrator (CI)	
65	tcp		tacacs-ds	TACACS-Database Service	
65		udp	tacacs-ds	TACACS-Database Service	
66	tcp		sql*net	Oracle SQL*NET	
66		udp	sql*net	Oracle SQL*NET	
67	tcp		bootps	Bootstrap Protocol Server	bootbs; boot protocol
67		udp	bootps	Bootstrap Protocol Server	bootbs; boot protocol
68	tcp		bootpc	Bootstrap Protocol Client	bootbs; boot protocol
68		udp	bootpc	Bootstrap Protocol Client	bootbs; boot protocol
69	tcp		tftp	Trivial File Transfer	tftp (trivial ftp); can update router s/w.
69		udp	tftp	Trivial File Transfer	tftp (trivial ftp); can update router s/w.
70	tcp		gopher	Gopher	gopher; outbound only
70		udp	gopher	Gopher	gopher; outbound only
71	tcp		netrjs-1	Remote Job Service	
71		udp	netrjs-1	Remote Job Service	
72	tcp		netrjs-2	Remote Job Service	
72		udp	netrjs-2	Remote Job Service	
73	tcp		netrjs-3	Remote Job Service	
73		udp	netrjs-3	Remote Job Service	
74	tcp		netrjs-4	Remote Job Service	
74		udp	netrjs-4	Remote Job Service	
75	tcp			any private dial out service	
75		udp		any private dial out service	
76	tcp			Unassigned	
76		udp		Unassigned	
77	tcp			any private RJE service	
77		udp		any private RJE service	
78	tcp		vettcp	vettcp	
78		udp	vettcp	vettcp	
79	tcp		finger	Finger	finger; who is logged in and where did it originate; outbound
79		udp	finger	Finger	finger; who is logged in and where did it originate; outbound
80	tcp		www	World Wide Web HTTP	http (hypertext transfer protocol); should be on proxy servers only
80		udp	www	World Wide Web HTTP	http (hypertext transfer protocol); should be on proxy servers only
81	tcp		hosts2-ns	HOSTS2 Name Server	
81		udp	hosts2-ns	HOSTS2 Name Server	
82	tcp		xfer	XFER Utility	
82		udp	xfer	XFER Utility	
83	tcp		mit-ml-dev	MIT ML Device	
83		udp	mit-ml-dev	MIT ML Device	
84	tcp		ctf	Common Trace Facility	

84		udp	ctf	Common Trace Facility	
85	tcp		mit-ml-dev	MIT ML Device	
85		udp	mit-ml-dev	MIT ML Device	
86	tcp		mfcobol	Micro Focus Cobol	
86		udp	mfcobol	Micro Focus Cobol	
87	tcp			any private terminal link	link
87		udp		any private terminal link	link
88	tcp		kerberos	Kerberos	kerberos
88		udp	kerberos	Kerberos	kerberos
89	tcp		su-mit-tg	SU/MIT Telnet Gateway	
89		udp	su-mit-tg	SU/MIT Telnet Gateway	
90	tcp		dnsix	DNSIX Securit Attribute Token Map	
90		udp	dnsix	DNSIX Securit Attribute Token Map	
91	tcp		mit-dov	MIT Dover Spooler	
91		udp	mit-dov	MIT Dover Spooler	
92	tcp		npp	Network Printing Protocol	
92		udp	npp	Network Printing Protocol	
93	tcp		dcp	Device Control Protocol	
93		udp	dcp	Device Control Protocol	
94	tcp		objcall	Tivoli Object Dispatcher	objcall; tivoli
94		udp	objcall	Tivoli Object Dispatcher	objcall; tivoli
95	tcp		supdup	SUPDUP	supdup
95		udp	supdup	SUPDUP	supdup
96	tcp		dixie	DIXIE Protocol Specification	
96		udp	dixie	DIXIE Protocol Specification	
97	tcp		swift-rvf	Swift Remote Vitural File Protocol	
97		udp	swift-rvf	Swift Remote Vitural File Protocol	
98	tcp		tacnews	TAC News	
98		udp	tacnews	TAC News	
99	tcp		metagram	Metagram Relay	
99		udp	metagram	Metagram Relay	
100	tcp		newacct	[unauthorized use]	
101	tcp		hostname	NIC Host Name Server	
101		udp	hostname	NIC Host Name Server	
102	tcp		iso-tsap	ISO-TSAP	
102		udp	iso-tsap	ISO-TSAP	
103	tcp		gppitnp	Genesis Point-to-Point Trans Net	
103		udp	gppitnp	Genesis Point-to-Point Trans Net	
104	tcp		acr-nema	ACR-NEMA Digital Imag. & Comm. 300	
104		udp	acr-nema	ACR-NEMA Digital Imag. & Comm. 300	
105	tcp		csnet-ns	Mailbox Name Nameserver	
105		udp	csnet-ns	Mailbox Name Nameserver	
106	tcp		3com-tsmux	3COM-TSMUX	
106		udp	3com-tsmux	3COM-TSMUX	
107	tcp		rtelnet	Remote Telnet Service	
107		udp	rtelnet	Remote Telnet Service	
108	tcp		snagas	SNA Gateway Access Server	

108		udp	snagas	SNA Gateway Access Server	
109	tcp		pop2	Post Office Protocol — Version 2	pop2
109		udp	pop2	Post Office Protocol —Version 2	pop2
110	tcp		pop3	Post Office Protocol — Version 3	POP3; needs a patch
110		udp	pop3	Post Office Protocol — Version 3	POP3; needs a patch
111	tcp		sunrpc	SUN Remote Procedure Call	rpc (remote procedure call); used to steal passwords and files
111		udp	sunrpc	SUN Remote Procedure Call	rpc (remote procedure call); used to steal passwords and files
112	tcp		mcidas	McIDAS Data Transmission Protocol	
112		udp	mcidas	McIDAS Data Transmission Protocol	
113	tcp		auth	Authentication Service	auth; limit
113		udp	auth	Authentication Service	auth; limit
114	tcp		audionews	Audio News Multicast	
114		udp	audionews	Audio News Multicast	
115	tcp		sftp	Simple File Transfer Protocol	
115		udp	sftp	Simple File Transfer Protocol	
116	tcp		ansanotify	ANSA REX Notify	
116		udp	ansanotify	ANSA REX Notify	
117	tcp		uucp-path	UUCP Path Service	
117		udp	uucp-path	UUCP Path Service	
118	tcp		sqlserv	SQL Services	
118		udp	sqlserv	SQL Services	
119	tcp		nntp	Network News Transfer Protocol	NNTP (Network News Transfer Protocol); assesses and reads news; may block
119		udp	nntp	Network News Transfer Protocol	NNTP (Network News Transfer Protocol); assesses and reads news; may block
120	tcp		cfdptkt	CFDPTKT	
120		udp	cfdptkt	CFDPTKT	
121	tcp		erpc	Encore Expedited Remote Pro.Call	erpc; encore
121		udp	erpc	Encore Expedited Remote Pro.Call	erpc; encore
122	tcp		smakynet	SMAKYNET	
122		udp	smakynet	SMAKYNET	
123	tcp		ntp	Network Time Protocol	ntp; network time protocol; may block
123		udp	ntp	Network Time Protocol	ntp; network time protocol; may block
124	tcp		ansatrader	ANSA REX Trader	
124		udp	ansatrader	ANSA REX Trader	
125	tcp		locus-map	Locus PC-Interface Net Map Ser	
125		udp	locus-map	Locus PC-Interface Net Map Ser	
126	tcp		unitary	Unisys Unitary Login	
126		udp	unitary	Unisys Unitary Login	
127	tcp		locus-con	Locus PC-Interface Conn Server	
127		udp	locus-con	Locus PC-Interface Conn Server	

128	tcp		gss-xlicen	GSS X License Verification	
128		udp	gss-xlicen	GSS X License Verification	
129	tcp		pwdgen	Password Generator Protocol	
129		udp	pwdgen	Password Generator Protocol	
130	tcp		cisco-fna	cisco FNATIVE	
130		udp	cisco-fna	cisco FNATIVE	
131	tcp		cisco-tna	cisco TNATIVE	
131		udp	cisco-tna	cisco TNATIVE	
132	tcp		cisco-sys	cisco SYSMAINT	
132		udp	cisco-sys	cisco SYSMAINT	
133	tcp		statsrv	Statistics Service	
133		udp	statsrv	Statistics Service	
134	tcp		ingres-net	INGRES-NET Service	
134		udp	ingres-net	INGRES-NET Service	
135	tcp		loc-srv	Location Service	location service; can cause NT to showq 100% cpu usage
135		udp	loc-srv	Location Service	location service; can cause NT to showq 100% cpu usage
136	tcp		profile	PROFILE Naming System	
136		udp	profile	PROFILE Naming System	
137	tcp		netbios-ns	NETBIOS Name Service	netbios, name service at router
137		udp	netbios-ns	NETBIOS Name Service	netbios, name service at router
138	tcp		netbios-dgm	NETBIOS Datagram Service	netbios, datagram service at router
138		udp	netbios-dgm	NETBIOS Datagram Service	netbios, datagram service at router
139	tcp		netbios-ssn	NETBIOS Session Service	netbios, session service at router
139		udp	netbios-ssn	NETBIOS Session Service	netbios, session service at router
140	tcp		emfis-data	EMFIS Data Service	
140		udp	emfis-data	EMFIS Data Service	
141	tcp		emfis-cntl	EMFIS Control Service	
141		udp	emfis-cntl	EMFIS Control Service	
142	tcp		bl-idm	Britton-Lee IDM	
142		udp	bl-idm	Britton-Lee IDM	
143	tcp		imap2	Interim Mail Access Protocol v2	IMAP; needs a patch
143		udp	imap2	Interim Mail Access Protocol v2	IMAP; needs a patch
144	tcp		news	NewS	
144		udp	news	NewS	
145	tcp		uaac	UAAC Protocol	
145		udp	uaac	UAAC Protocol	
146	tcp		iso-tp0	ISO-IP0	
146		udp	iso-tp0	ISO-IP0	
147	tcp		iso-ip	ISO-IP	
147		udp	iso-ip	ISO-IP	
148	tcp		cronus	CRONUS-SUPPORT	
148		udp	cronus	CRONUS-SUPPORT	
149	tcp		aed-512	AED 512 Emulation Service	
149		udp	aed-512	AED 512 Emulation Service	
150	tcp		sql-net	SQL-NET	

150		udp	sql-net	SQL-NET	
151	tcp		hems	HEMS	
151		udp	hems	HEMS	
152	tcp		bftp	Background File Transfer Program	
152		udp	bftp	Background File Transfer Program	
153	tcp		sgmp	SGMP	
153		udp	sgmp	SGMP	
154	tcp		netsc-prod	NETSC	
154		udp	netsc-prod	NETSC	
155	tcp		netsc-dev	NETSC	
155		udp	netsc-dev	NETSC	
156	tcp		sqlsrv	SQL Service	
156		udp	sqlsrv	SQL Service	
157	tcp		knet-cmp	KNET/VM Command/Message Protocol	
157		udp	knet-cmp	KNET/VM Command/Message Protocol	
158	tcp		pcmail-srv	PCMail Server	
158		udp	pcmail-srv	PCMail Server	
159	tcp		nss-routing	NSS-Routing	
159		udp	nss-routing	NSS-Routing	
160	tcp		sgmp-traps	SGMP-TRAPS	
160		udp	sgmp-traps	SGMP-TRAPS	
161	tcp		snmp	SNMP	
161		udp	snmp	SNMP	
162	tcp		snmptrap	SNMPTRAP	
162		udp	snmptrap	SNMPTRAP	
163	tcp		cmip-man	CMIP/TCP Manager	
163		udp	cmip-man	CMIP/TCP Manager	
164	tcp		cmip-agent	CMIP/TCP Agent	
164		udp	smip-agent	CMIP/TCP Agent	
165	tcp		xns-courier	Xerox	
165		udp	xns-courier	Xerox	
166	tcp		s-net	Sirius Systems	
166		udp	s-net	Sirius Systems	
167	tcp		namp	NAMP	
167		udp	namp	NAMP	
168	tcp		rsvd	RSVD	
168		udp	rsvd	RSVD	
169	tcp		send	SEND	
169		udp	send	SEND	
170	tcp		print-srv	Network PostScript	
170		udp	print-srv	Network PostScript	
171	tcp		multiplex	Network Innovations Multiplex	
171		udp	multiplex	Network Innovations Multiplex	
172	tcp		cl/1	Network Innovations CL/1	
172		udp	cl/1	Network Innovations CL/1	
173	tcp		xyplex-mux	Xyplex	
173		udp	xyplex-mux	Xyplex	
174	tcp		mailq	MAILQ	

174		udp	mailq	MAILQ	
175	tcp		vmnet	VMNET	
175		udp	vmnet	VMNET	
176	tcp		genrad-mux	GENRAD-MUX	
176		udp	genrad-mux	GENRAD-MUX	
177	tcp		xdmcp	X Display Manager Control Protocol	
177		udp	xdmcp	X Display Manager Control Protocol	
178	tcp		nextstep	NextStep Window Server	
178		udp	NextStep	NextStep Window Server	
179	tcp		bgp	Border Gateway Protocol	
179		udp	bgp	Border Gateway Protocol	
180	tcp		ris	Intergraph	
180		udp	ris	Intergraph	
181	tcp		unify	Unify	
181		udp	unify	Unify	
182	tcp		audit	Unisys Audit SITP	
182		udp	audit	Unisys Audit SITP	
183	tcp		ocbinder	OCBinder	
183		udp	ocbinder	OCBinder	
184	tcp		ocserver	OCServer	
184		udp	ocserver	OCServer	
185	tcp		remote-kis	Remote-KIS	
185		udp	remote-kis	Remote-KIS	
186	tcp		kis	KIS Protocol	
186		udp	kis	KIS Protocol	
187	tcp		aci	Application Communication Interface	
187		udp	aci	Application Communication Interface	
188	tcp		mumps	Plus Five's MUMPS	
188		udp	mumps	Plus Five's MUMPS	
189	tcp		qft	Queued File Transport	
189		udp	qft	Queued File Transport	
190	tcp		gacp	Gateway Access Control Protocol	
190		udp	cacp	Gateway Access Control Protocol	
191	tcp		prospero	Prospero	
191		udp	prospero	Prospero	
192	tcp		osu-nms	OSU Network Monitoring System	
192		udp	osu-nms	OSU Network Monitoring System	
193	tcp		srmp	Spider Remote Monitoring Protocol	
193		udp	srmp	Spider Remote Monitoring Protocol	
194	tcp		irc	Internet Relay Chat Protocol	
194		udp	irc	Internet Relay Chat Protocol	
195	tcp		dn6-nlm-aud	DNSIX Network Level Module Audit	

195		udp	dn6-nlm-aud	DNSIX Network Level Module Audit	
196	tcp		dn6-smm-red	DNSIX Session Mgt Module Audit Redir	
196		udp	dn6-smm-red	DNSIX Session Mgt Module Audit Redir	
197	tcp		dls	Directory Location Service	
197		udp	dls	Directory Location Service	
198	tcp		dls-mon	Directory Location Service Monitor	
198		udp	dls-mon	Directory Location Service Monitor	
199	tcp		smux	SMUX	
199		udp	smux	SMUX	
200	tcp		src	IBM System Resource Controller	
200		udp	src	IBM System Resource Controller	
201	tcp		at-rtmp	AppleTalk Routing Maintenance	
201		udp	at-rtmp	AppleTalk Routing Maintenance	
202	tcp		at-nbp	AppleTalk Name Binding	
202		udp	at-nbp	AppleTalk Name Binding	
203	tcp		at-3	AppleTalk Unused	
203		udp	at-3	AppleTalk Unused	
204	tcp		at-echo	AppleTalk Echo	
204		udp	at-echo	AppleTalk Echo	
205	tcp		at-5	AppleTalk Unused	
205		udp	at-5	AppleTalk Unused	
206	tcp		at-zis	AppleTalk Zone Information	
206		udp	at-zis	AppleTalk Zone Information	
207	tcp		at-7	AppleTalk Unused	
207		udp	at-7	AppleTalk Unused	
208	tcp		at-8	AppleTalk Unused	
208		udp	at-8	AppleTalk Unused	
209	tcp		tam	Trivial Authenticated Mail Protocol	
209		udp	tam	Trivial Authenticated Mail Protocol	
210	tcp		z39.50	ANSI Z39.50	
210		udp	z39.50	ANSI Z39.50	
211	tcp		914c/g	Texas Instruments 914C/G Terminal	
211		udp	914c/g	Texas Instruments 914C/G Terminal	
212	tcp		anet	ATEXSSTR	
212		udp	anet	ATEXSSTR	
213	tcp		ipx	IPX	
213		udp	ipx	IPX	
214	tcp		vmpwscs	VM PWSCS	
214		udp	vmpwscs	VM PWSCS	
215	tcp		softpc	Insignia Solutions	
215		udp	softpc	Insignia Solutions	

216	tcp		atls	Access Technology License Server	
216		udp	atls	Access Technology License Server	
217	tcp		dbase	dBASE UNIX	
217		udp	dbase	dBASE UNIX	
218	tcp		mpp	Netix Message Posting Protocol	
218		udp	mpp	Netix Message Posting Protocol	
219	tcp		uarps	Unisys ARPs	
219		udp	uarps	Unisys ARPs	
220	tcp		imap3	Interactive Mail Access Protocol v3	
220		udp	imap3	Interactive Mail Access Protocol v3	
221	tcp		fln-spx	Berkeley rlogind with SPX auth	
221		udp	fln-spx	Berkeley rlogind with SPX auth	
222	tcp		fsh-spx	Berkeley rshd with SPX auth	
222		udp	fsh-spx	Berkeley rshd with SPX auth	
223	tcp		cdc	Certificate Distribution Center	
223		udp	cdc	Certificate Distribution Center	
224-241	-	-		Reserved	
243	tcp		sur-meas	Survey Measurement	
243		udp	sur-meas	Survey Measurement	
245	tcp		link	LINK	
245		udp	link	LINK	
246	tcp		dsp3270	Display Systems Protocol	
246		udp	dsp3270	Display Systems Protocol	
247-255	-	-		Reserved	
345	tcp		pawserv	Perf Analysis Workbench	
345		udp	pawserv	Perf Analysis Workbench	
346	tcp		zserv	Zebra server	
346		udp	zserv	Zebra server	
347	tcp		fatserv	Fatmen Server	
347		udp	fatserv	Fatmen Server	
371	tcp		clearcase	Clearcase	
371		udp	clearcase	Clearcase	
372	tcp		ulistserv	UNIX Listserv	
372		udp	ulistserv	UNIX Listserv	
373	tcp		legent-1	Legent Corporation	
373		udp	legent-1	Legent Corporation	
374	tcp		legent-2	Legent Corporation	
374		udp	legent-2	Legent Corporation	
512	tcp		exec	remote process execution; authentication performed using passwords and UNIX login names	rxe (remote execute)
512		udp	biff	used by mail system to notify users of new mail received; currently receives messages only from processes on the same machine	rxe (remote execute)

513	tcp		login	remote login via telnet; automatic authentication performed based on priviledged port numbers and distributed data bases which identify "authentication domains"	rlogin (remote login)
513		udp	who	maintains data bases showing who's logged into machines on a local net and the load average of the machine	rlogin (remote login)
514	tcp		cmd	like exec, but automatic authentication is performed as for login server	rsh (remote shell)
514		udp	syslog		rsh (remote shell)
515	tcp		printer	spooler	
515		udp	printer	spooler	
517	tcp		talk	like tenex link, but across machine — unfortunately, doesn't use link protocol (this is actually just a rendezvous port from which a tcp connection is established)	
517		udp	talk	like tenex link, but across machine — unfortunately, doesn't use link protocol (this is actually just a rendezvous port from which a tcp connection is established)	
518	tcp		ntalk		
518		udp	ntalk		
519	tcp		utime	UNIXtime	
519		udp	utime	UNIXtime	
520	tcp		efs	extended file name server	RIP (Routing Information Protocol); can be spoofed
520		udp	router	local routing process (on site); uses variant of Xerox NS routing information protocol	RIP (Routing Information Protocol); can be spoofed
525	tcp		timed	timeserver	
525		udp	timed	timeserver	
526	tcp		tempo	newdate	
526		udp	tempo	newdate	
530	tcp		courier	rpc	
530		udp	courier	rpc	
531	tcp		conference	chat	
531		udp	conference	chat	
532	tcp		netnews	readnews	
532		udp	netnews	readnews	
533	tcp		netwall	for emergency broadcasts	
533		udp	netwall	for emergency broadcasts	
540	tcp		uucp	uucpd	UUCP (UNIX to UNIX copy); can give unauthorized access if misconfigured

540		udp	uucp	uucpd	UUCP (UNIX to UNIX copy); can give unauthorized access if misconfigured
543	tcp		klogin		
543		udp	klogin		
544	tcp		kshell	krcmd	
544		udp	kshell	krcmd	
550	tcp		new-rwho	new-who	
550		udp	new-rwho	new-who	
555	tcp		dsf		
555		udp	dsf		
556	tcp		remotefs	rfs server	
556		udp	remotefs	rfs server	
560	tcp		rmonitor	rmonitord	
560		udp	rmonitor	rmonitord	
561	tcp		monitor		
561		udp	monitor		
562	tcp		chshell	chcmd	
562		udp	chshell	chcmd	
564	tcp		9pfs	plan 9 file service	
564		udp	9pfs	plan 9 file service	
565	tcp		whoami	whoami	
565		udp	whoami	whoami	
570	tcp		meter	demon	
570		udp	meter	demon	
571	tcp		meter	udemon	
571		udp	meter	udemon	
600	tcp		ipcserver	Sun IPC server	
600		udp	ipcserver	Sun IPC server	
607	tcp		nqs	nqs	
607		udp	nqs	nqs	
666	tcp		mdqs		
666		udp	mdqs		
704	tcp		elcsd	errlog copy/server daemon	
704		udp	elcsd	errlog copy/server daemon	
740	tcp		netcp	NETscout Control Protocol	
740		udp	netcp	NETscout Control Protocol	
741	tcp		netgw	netGW	
741		udp	netgw	netGW	
742	tcp		netrcs	Network based Rev. Cont. Sys.	
742		udp	netrcs	Network based Rev. Cont. Sys.	
744	tcp		flexlm	Flexible License Manager	flexn
744		udp	flexlm	Flexible License Manager	flexn
747	tcp		fujitsu-dev	Fujitsu Device Control	
747		udp	fujitsu-dev	Fujitsu Device Control	
748	tcp		ris-cm	Russell Info Sci Calendar Manager	
748		udp	ris-cm	Russell Info Sci Calendar Manager	
749	tcp		kerberos-adm	kerberos administration	
749		udp	kerberos-adm	kerberos administration	
750	tcp		rfile		
750		udp	loadav		

751	tcp		pump		
751		udp	pump		
752	tcp		qrh		
752		udp	qrh		
753	tcp		rrh		
753		udp	rrh		
754	tcp		tell	send	
754		udp	tell	send	
758	tcp		nlogin		
758		udp	nlogin		
759	tcp		con		
759		udp	con		
760	tcp		ns		
760		udp	ns		
761	tcp		rxe		
761		udp	rxe		
762	tcp		quotad		
762		udp	quotad		
763	tcp		cycleserv		
763		udp	cycleserv		
764	tcp		omserv		
764		udp	omserv		
765	tcp		webster		
765		udp	webster		
767	tcp		phonebook	phone	
767		udp	phonebook	phone	
769	tcp		vid		
769		udp	vid		
770	tcp		cadlock		
770		udp	cadlock		
771	tcp		rtip		
771		udp	rtip		
772	tcp		cycleserv2		
772		udp	cycleserv2		
773	tcp		submit		
773		udp	notify		
774	tcp		rpasswd		
774		udp	acmaint_dbd		
775	tcp		entomb		
775		udp	acmaint_transd		
776	tcp		wpages		
776		udp	wpages		
780	tcp		wpgs		
780		udp	wpgs		
781	tcp		hp-collector	hp performance data collector	
781		udp	hp-collector	hp performance data collector	
782	tcp		hp-managed-node	hp performance data managed node	
782		udp	hp-managed-node	hp performance data managed node	
783	tcp		hp-alarm-mgr	hp performance data alarm manager	

783		udp	hp-alarm-mgr	hp performance data alarm manager	
800	tcp		mdbs_daemon		
800		udp	mdbs_daemon		
801	tcp		device		
801		udp	device		
996	tcp		xtreelic	XTREE License Server	
996		udp	xtreelic	XTREE License Server	
997	tcp		maitrd		
997		udp	maitrd		
998	tcp		busboy		
998		udp	puparp		
999	tcp		garcon		
999		udp	applix	Applix ac	
999	tcp		puprouter		
999		udp	puprouter		
1000	tcp		cadlock		
1000		udp	ock		
1025	tcp		blackjack	network blackjack	
1025		udp	blackjack	network blackjack	
1248	tcp		hermes		
1248		udp	hermes		
1347	tcp		bbn-mmc	multi media conferencing	
1347		udp	bbn-mmc	multi media conferencing	
1348	tcp		bbn-mmx	multi media conferencing	
1348		udp	bbn-mmx	multi media conferencing	
1349	tcp		sbook	Registration Network Protocol	
1349		udp	sbook	Registration Network Protocol	
1350	tcp		editbench	Registration Network Protocol	
1350		udp	editbench	Registration Network Protocol	
1351	tcp		equationbuilder	Digital Tool Works (MIT)	
1351		udp	equationbuilder	Digital Tool Works (MIT)	
1352	tcp		lotusnote	Lotus Note	
1352		udp	lotusnote	Lotus Note	
1524	tcp		ingreslock	ingres	
1524		udp	ingreslock	ingres	
1525	tcp		orasrv	oracle	
1525		udp	orasrv	oracle	
1525	tcp		prospero-np	prospero nonprivileged	
1525		udp	prospero-np	prospero nonprivileged	
1527	tcp		tlisrv	oracle	
1527		udp	tlisrv	oracle	
1529	tcp		coauthor	oracle	
1529		udp	coauthor	oracle	
1600	tcp		issd		
1600		udp	issd		
1650	tcp		nkd		
1650		udp	nkd		
2000	tcp		callbook		OpenWindows; displays data and keystroke s

2000		udp	callbook		OpenWindows; displays data and keystrokes
2001	tcp		dc		
2001		udp	wizard	curry	
2002	tcp		globe		
2002		udp	globe		
2004	tcp		mailbox		
2004		udp	emce	CCWS mm conf	
2005	tcp		berknet		
2005		udp	oracle		
2006	tcp		invokator		
2006		udp	raid-cc	raid	
2007	tcp		dectalk		
2007		udp	raid-am		
2008	tcp		conf		
2008		udp	terminaldb		
2009	tcp		news		
2009		udp	whosockami		
2010	tcp		search		
2010		udp	pipe_server		
2011	tcp		raid-cc	raid	
2011		udp	servserv		
2012	tcp		ttyinfo		
2012		udp	raid-ac		
2013	tcp		raid-am		
2013		udp	raid-cd		
2014	tcp		troff		
2014		udp	raid-sf		
2015	tcp		cypress		
2015		udp	raid-cs		
2016	tcp		bootserver		
2016		udp	bootserver		
2017	tcp		cypress-stat		
2017		udp	bootclient		
2018	tcp		terminaldb		
2018		udp	rellpack		
2019	tcp		whosockami		
2019		udp	about		
2020	tcp		xinupageserver		
2020		udp	xinupageserver		
2021	tcp		servexec		
2021		udp	xinuexpansion		
2022	tcp		down		
22022		udp	xinuexpansion		
32023	tcp		xinuexpansion		
32023		udp	xinuexpansion		
42024	tcp		xinuexpansion		
42024		udp	xinuexpansion		
2025	tcp		ellpack		
2025		udp	xribs		

2026	tcp		scrabble		
2026		udp	scrabble		
2027	tcp		shadowserver		
2027		udp	shadowserver		
2028	tcp		submitserver		
2028		udp	submitserver		
2030	tcp		device2		
2030		udp	device2		
2032	tcp		blackboard		
2032		udp	blackboard		
2033	tcp		glogger		
2033		udp	glogger		
2034	tcp		scoremgr		
2034		udp	scoremgr		
2035	tcp		imsldoc		
2035		udp	imsldoc		
2038	tcp		objectmanager		
2038		udp	objectmanager		
2040	tcp		lam		
2040		udp	lam		
2041	tcp		interbase		
2041		udp	interbase		
2042	tcp		isis		
2042		udp	isis		
2043	tcp		isis-bcast		
2043		udp	isis-bcast		
2044	tcp		rimsl		
2044		udp	rimsl		
2045	tcp		cdfunc		
2045		udp	cdfunc		
2046	tcp		sdfunc		
2046		udp	sdfunc		
2047	tcp		dls		
2047		udp	dls		
2048	tcp		dls-monitor		
2048		udp	dls-monitor		
2049	tcp		shilp		
2049		udp	shilp		
2766	tcp				
2766		udp			
2784	tcp		www-dev	World Wide Web - development	
2784		udp	www-dev	World Wide Web - development	
3049	tcp		NSWS		
3049		udp	NSWS		
3264	tcp				
3264		udp			
4672	tcp		rfa	remote file access server	
4672		udp	rfa	remote file access server	
5000	tcp		commplex-main		
5000		udp	commplex-main		
5001	tcp		commplex-link		

5001		udp	commplex-link		
5002	tcp		rfe	radio free ethernet	
5002		udp	rfe	radio free ethernet	
5145	tcp		rmonitor_secure		
5145		udp	rmonitor_secure		
5236	tcp		padl2sim		
5236		udp	padl2sim		
6000-6063	-	-	X-Windows	X-Windows	displays data and keystrokes
6111	tcp		sub-process	HP SoftBench Sub-Process Control	
6111		udp	sub-process	HP SoftBench Sub-Process Control	
6558		udp	xdsxdm		
6558	tcp		xdsxdm		
7000	tcp		afs3-fileserver	file server itself	
7000		udp	afs3-fileserver	file server itself	
7001	tcp		afs3-callback	callbacks to cache managers	
7001		udp	afs3-callback	callbacks to cache managers	
7002	tcp		afs3-prserver	users & groups database	
7002		udp	afs3-prserver	users & groups database	
7003	tcp		afs3-vlserver	volume location database	
7003		udp	afs3-vlserver	volume location database	
7004	tcp		afs3-kaserver	AFS/Kerberos authentication service	
7004		udp	afs3-kaserver	AFS/Kerberos authentication service	
7005	tcp		afs3-volser	volume managment server	
7005		udp	afs3-volser	volume managment server	
7006	tcp		afs3-errors	error interpretation service	
7006		udp	afs3-errors	error interpretation service	
7007	tcp		afs3-bos	basic overseer process	
7007		udp	afs3-bos	basic overseer process	
7008	tcp		afs3-update	server-to-server updater	
7008		udp	afs3-update	server-to-server updater	
7009	tcp		afs3-rmtsys	remote cache manager service	
7009		udp	afs3-rmtsys	remote cache manager service	
9535	tcp		man		
9535		udp	man		
17007	tcp		isode-dua		
17007		udp	isode-dua		

Appendix F — Construction Specifications Guideline

PROJECT: ______________________________ **DATE:** __________

What contract is used between the Owner and Contractor? ( ) AIA A101, ( ) EJCDC 1910-8-A-1, ( ) other ______________________________

Who will provide the General Conditions of the Contract?

( ) Consultant or ( ) Owner, using: ( ) AIA A210 ( ) other ______________________________

Will the Contract have any supplemental conditions? ______________________________

Is there any bonding requirements? ( ) Bid Bond ( ) Performance Bond ( ) Payment Bond
Amount of Bonds: ______________________________

What are the liability insurance requirements for the Contractor? ______________________________

What type of specifications will be used? ( ) Performance ( ) Open Proprietary ( ) Closed Proprietary
( ) Descriptive ( ) Combination

Is there any preference toward brands of equipment? ______________________________
____

Is there a requirement to match existing equipment? ______________________________
____

Will submittals be required if bidding specified equipment? ______________________________

How many copies of the submittals? ______________________________

Will samples/equipment demonstrations be required? ______________________________

How long should the warranty period be? ______________________________

Will an extended service contract be asked for? ______________________________

Will training be performed to: ( ) one group ( ) several groups ______________________________

Will any items need to be "paint to match?" ______________________________

Does the Authority Having Jurisdiction require any plan review? ______________________________

Will in-house crew provide any support services? ______________________________

Will there be any storage space provided for the contractor? ______________________________

Will any spare parts be required? ______________________________

Will the Project Manual be produced using ( ) single-sided or ( ) double-sided printing?

Appendix G — System Integration Checklist

Systems to Integrate	Design Elements	Regulatory Agencies
❒ Access Control	❒ Turnkey/System Integrator	❒ UL - Underwriters Lab
❒ CCTV System	❒ Sub-con. to General Contractor	❒ FM - Factory Mutual
❒ Fire Alarm/Life Safety	❒ Consultant/Design Professional	❒ BOCA - Building Officials
❒ Fuel Management	❒ New Construction Area	❒ ISO - Insurance Services
❒ Intercom/Paging	❒ Retrofit of Existing Devices	❒ IRI - Industrial Relations
❒ Intrusion Detection	❒ Schedules/Milestones	❒ DOD - Dept. of Defense
❒ Photo Imaging	❒ Pre-Qualify Contractors	❒ Local License & Public Acts
❒ Time & Attendance	❒ Card Technology ____________	❒ State License & Public Acts
Communication Medias	**Installation Items**	**Approvals & Listings**
❒ Hardwired	❒ Off-the-Shelf Products	❒ UL294 - Access Control
❒ Fiber Optics	❒ Interface to Existing Systems	❒ UL684 - Fire Alarm
❒ Broadband Coaxial	❒ Interface to New Systems	❒ UL983 - CCTV
❒ Microwave	❒ Factory Acceptance Testing	❒ UL1076 - Security
❒ Radio	❒ Custom Consoles	❒ NFPA72 - Fire Codes
❒ Leased Lines (ISDN)	❒ Shop Drawings	❒ NICET - Fire Certification
❒ Dial-up Telephone	❒ Service & Maintenance	❒ ISO9000
❒ Data Networks	❒ As-built Drawings	❒ FCC Federal Approval
Training Requirements	**System Capacities**	**Software Features**
❒ On-site Classes	Card Holders # ________________	❒ Single User
❒ Factory Classes	Card Readers # ________________	❒ Multi-Users
❒ Course Outline & Manual	Access Groups # _______________	❒ Multi-Tasking
❒ Video Taping	Time Zones # __________________	❒ Color Graphics
❒ Operator Classes	Alarm Points # ________________	❒ Networks (LANS/WANS)
❒ Maintenance Classes	Remote Sites # ________________	❒ Integ. Badging & Database
❒ O/M Manuals	Workstations # ________________	❒ Bar Coding
❒ Spare Parts	❒ MIS Interfaces	❒ Map Display

Additional Comments/ ____

Source: April 95 issue of Security Technology & Design. Forwarded by Fred Zagurski.

Appendix H — Individual Door/Gate Access Checklist

PROJECT: ______________________________ **DATE:** __________

Door/Gate Number: ______ Location: ______________________________

A. ***Frequency of Use & Risk***

Passages per day: ________ week: __________ Reader Vandalism Risk: ( ) Low ( ) Medium ( ) High

B. ***Door Type:***

( ) Steel ( ) Solid Wood ( ) Hollow Core ( ) Aluminum Frame Glass ( ) Solid Glass ( ) Fire Door

C. ***Door Frame:***

( ) Wood ( ) Aluminum ( ) Steel ( ) Concrete-filled Steel

D. ***Door Movement:***

( ) LHH ( ) RHH ( ) LHHR ( ) RHHR ( ) Double with fixed post ( ) Double with removable post
( ) Double with lip ( ) Door Closure ( ) Required Fire Exit

E. ***Existing Mechanical Hardware***

( ) Cylindrical Lock ( ) Mortise Lock ( ) Rim Lock ( ) Deadbolt Lock ( ) Mechanical Exit Bar

F. ***Proposed Electric Door Lock***

( ) Electric Strike ( ) Electric Deadbolt ( ) Electric Mortise Lock ( ) Electromagnet ( ) Electric Hinge
( ) Fire-Safe ( ) Fire-Secure Voltage ________ Power Supply location ______________________

G. ***Gate Type***: ( ) LIFT ( ) PIVOT ( ) SWING ( ) SLIDE [ ] V-TRACK [ ] CANTILEVER [ ] OTHER ________

H. ***Gate Safety Device:*** ( ) LOOP DETECTOR ( ) PHOTOELECTRIC EYE ( ) EDGE SENSOR ( ) AUTO-REBOUND

I. ***Gate Operating Power:*** __________ VOLTS, _____ PHASE, ( ) PROVIDED BY OWNER ( ) INSTALL NEW

J. ***Door/Gate Status Sensor:***

( ) Magnetic Contact ( ) Integral with Lock ( ) Latchbolt Monitor ( ) Other ____________________

K. ***Door/Gate Control***

Free Access during

Free Egress: ( ) 24 hours ( ) During ______________________________

Door Egress Method: ( ) Reader ( ) Motion ( ) Button ( ) TSB ( ) Phone ( ) Intercom ( ) Guard

Gate Egress Method: ( ) Reader ( ) Beam ( ) Push Button ( ) Buried Loop ( ) Radio ( ) Guard

Fire Alarm Action: ( ) Free Egress ( ) Free Access ( ) Delayed Egress ________________ seconds
Auxiliary Function: ( ) Time Attendance ( ) Guard Tour ( ) other ______________________
Controller Location: ______________________________

Appendix I — Retrofit Versus New Construction

New Construction	Retrofit
Formal specifications and drawings	Security contractor offers proposal (list of equipment)
General contractor in charge — security under/with electrical	Security contractor in charge — may subcontract to electrical
Conduit and sometimes wire provided by electrical contractor	All equipment by security contractor
Requires much vigilance for subcontractor coordination	All work coordinated by security contractor
Interface to electrical boxes required	Equipment is attached directly to walls, doors, etc.
All work comes under electrical inspection	Work may not be inspected
Work may be listed under several sections	All work comes under one section
Drawings almost always required	Drawings seldom requested
Full documentation required	Operating manual usually provided
Full training required (sometimes at factory)	Minimal training provided
Spare parts often requested	Spare parts seldom offered
Service response time often required	Service time may be within 24 hours under contract
License, insurance, and bonding required	License and insurance may be requested
Security contractor trains user how to program system	Security contractor programs system
Security wiring is labeled and a list provided to user	Wiring may be labeled for reference, no list is provided
System circuit readings are required and given to user	System circuit readings are not taken
Complete inspection of the system is required	Security contractor may make a random inspection
Complete testing of the system is performed	Security contractor may make a random test of the system

This is a legal agreement between you (either an individual or and entity), the end user, and Butterworth-Heinemann. If you do not agree to the terms of this Agreement, promptly return the disk and the accompanying manual to Butterworth-Heinemann.

Butterworth-Heinemann Software Licensing Agreement

1. GRANT OF LICENSE. This Butterworth-Heinemann License Agreement ("License") permits you to use one copy of the specified version of the Butterworth-Heinemann software product identified above ("Software") on any single computer, provided the SOFTWARE is in use on only one computer at any time. If you have multiple Licenses for the SOFTWARE, then at any time you may have as many copies of the SOFTWARE in use as you have Licenses. The SOFTWARE is "in use" on a computer when it is loaded into the temporary memory (i.e., RAM) or installed into the permanent memory (e.g., hard disk, CD-ROM, or other storage device) of that computer, except for a copy installed on a network server for the sole purpose of distribution to other computers is not "in use." If the anticipated number of users of the SOFTWARE will exceed the number of applicable Licenses, then you must have a reasonable mechanism or process in place to assure that the number of persons using the SOFTWARE concurrently does not exceed the number of Licenses. If the SOFTWARE is permanently installed on the hard disk or other storage device of a computer (other than a network server) and one person uses that computer more than 80% of the time it is in use, then that person may also use the SOFTWARE on a portable or home computer.

2. COPYRIGHT. The SOFTWARE is owned by Butterworth-Heinemann or its suppliers and is protected by United States copyright laws and international treaty provisions. Therefore, you must treat the SOFTWARE like any other copyrighted material (e.g., a book or musical recording) EXCEPT that you may either (a) make one copy of the SOFTWARE solely for backup or archival purposes, (b) transfer the SOFTWARE to a single hard disk provided you keep the original solely for backup or archival purposes, or (c) produce a survey from the questions contained in *The Ultimate Security Survey, Second Edition* on disk that is solely used in-house. This software file and its component questions and material may not be reproduced for or distributed to any other party and may not be sold in any form. You may not copy the written materials accompanying the SOFTWARE.

3. OTHER RESTRICTIONS. This Butterworth-Heinemann License Agreement is your proof of license to exercise the rights granted herein and must be retained by you. You may not rent or lease the SOFTWARE and you may not transfer your rights under this Butterworth-Heinemann License Agreement on any basis to another individual or entity.

4. OTHER RESTRICTIONS. *The Ultimate Security Survey, Second Edition*, Copyright © 1998 Butterworth-Heinemann, is intended for personal use only. Reproduction of any part of this file for distribution or sale to any other party is strictly prohibited.

U.S. GOVERNMENT RESTRICTED RIGHTS

The SOFTWARE and documentation are provided with RESTRICTED RIGHTS. Use, duplication, or disclosure by the Government is subject to restrictions as set fourth in subparagraph (c)(1)(ii) of the Rights in Technical Data and computer Software clause at DFARS 252.227-7013 or subparagraphs (c)(1) and (2) of the Commercial Computer Software—Restricted Rights at 48 CFR 52.227-19, as applicable. Contractor/manufacturer is Butterworth-Heinemann, 225 Wildwood Avenue, Woburn, MA 01801-2041.

This Agreement is governed by the laws of the State of Massachusetts.

The Ultimate Security Survey, Second Edition

Disk Instructions

The included disk provides all questions contained in *The Ultimate Security Survey, Second Edition*. The file is formatted in standard "TXT" (ASCII text) format. You can easily design surveys with your own IBM-compatible word processing system or on a Macintosh (with the appropriate conversion software).

To use the disk material, insert the disk into the 3.5-inch drive of your computer while your word processing system is running.

Open the file ULTIMV2.TXT as you would any other word processing file. ULTIMV2.TXT is best viewed with the following settings: Top 1", Bottom 1", Left 1", and Right 1". Remember to switch to the appropriate drive, and ask the system to list files with the extension "TXT". Your computer may prompt you as to whether you want to convert the file to your current word processor format—you will want to. The word processor should do the rest.

After you've converted and copied the file, write protect the disk by sliding the plastic square in the top back corner of the disk until it snaps into place. A small square hold will appear in the corner. This will prevent you from over-writing the disk and erasing any material you will want to use again.

The Ultimate Security Survey, Second Edition on disk is designed to work in conjunction with the text of *The Ultimate Security Survey, Second Edition*. Please review *The Ultimate Security Survey, Second Edition* and circle the questions you would like included in your survey. On the disk version of *The Ultimate Security Survey, Second Edition*, copy (cut and paste) the corresponding questions into your word processor; you will then have your completed security survey.